# An Atoms First Approach to the General Chemistry Laboratory

## Second Edition

Gregg R. Dieckmann
*The University of Texas at Dallas*

John W. Sibert
*The University of Texas at Dallas*

Mc
Graw
Hill
Education

AN ATOMS FIRST APPROACH TO THE GENERAL CHEMISTRY LABORATORY, SECOND EDITION

5 6 7 8 9 10 QVS/QVS 21 20 19 18 17

ISBN 978-0-07-764642-4
MHID 0-07-764642-8

Senior Vice President, Products & Markets: *Kurt L. Strand*
Vice President, General Manager, Products & Markets: *Marty Lange*
Vice President, Content Production & Technology Services: *Kimberly Meriwether David*
Managing Director: *Thomas Timp*
Executive Brand Manager: *David Spurgeon, Ph.D.*
Director of Development: *Rose M. Koos*
Senior Product Developer: *Mary E. Hurley*
Director of Digital Content: *Shirley Hino, Ph.D.*
Digital Product Analyst: *Patrick Diller*
Executive Marketing Manager: *Tamara L. Hodge*
Director, Content Production: *Terri Schiesl*
Content Project Manager (Print): *Sheila M. Frank*
Content Project Manager (Media): *Daryl Bruflodt*
Senior Buyer: *Laura Fuller*
Senior Designer: *David W. Hash*
Cover Image: *Artistic representation of high quality graphene—a "wonder" material that is the thinnest and strongest in the world. Kaustav Banerjee, professor in Electrical and Computer Engineering and Director of the Nanoelectronics Research Lab at UCSB, led the research team to perfect methods of growing sheets of graphene. ©Peter Allen UC Santa Barbara.*
Content Licensing Specialist: *Shawntel Schmitt*
Compositor: *Aptara,® Inc.*
Typeface: *11/13 Times*
Printer: *Quad/Graphics*

# Contents

# Preface

## OVERVIEW

This laboratory manual presents a curriculum that is organized around an *atoms first* approach to general chemistry. Our motivation for writing this manual is to (1) tap into the natural curiosity present in all of us and provide engaging experiments that students will find interesting, (2) emphasize topics that students find particularly challenging in the general chemistry lecture course, and (3) create a laboratory environment that encourages students, on occasion, to "solve puzzles" and not just "follow recipes." All too often, students view general chemistry lab as a boring exercise in which an exact set of instructions is followed, leading to an answer that, in many cases, results in a good grade regardless of how much learning has taken place. To these students, the successful lab is the one that takes the least amount of time! Unfortunately, a huge opportunity to get students truly turned on to science is missed. To us, the laboratory represents high-stakes ground for engagement and relatively low stakes for grading, as the laboratory is typically a single-credit course or minor component to the lecture grade. Thus, while the rigor of the experiments in this manual can be tuned to meet the needs of the instructor, our hope is that students will be encouraged to "play" (safely) with chemical concepts and laboratory techniques, with grades simply being a natural consequence of their laboratory actions. To facilitate such a mindset, this manual has been written to provide instructors with a weekly tool that can attract and keep student interest, while providing important connections to the material covered in an atoms first lecture course. Our philosophy: *student curiosity* leads to *engagement*, which leads to *discovery*, which leads to *learning*.

The manual is for a freshman-level general chemistry laboratory course, and serves as an ideal supplement for any atoms first general chemistry textbook (such as *Chemistry: Atoms First* by Julia Burdge and Jason Overby). It is designed for students at all levels, from those seeing chemistry for the first time to chemistry majors.

## IMPORTANT FEATURES OF THIS LABORATORY MANUAL

- Twenty-five self-contained experiments; early experiments focus on topics introduced at the beginning of an atoms first course—properties of light and the use of light to study nanomaterials, line spectra and the structure of atoms, and periodic trends.
- We recommend that prelab exercises for each experiment be cast in the form of student workshops— group activities based on a peer-lead-team-learning (PLTL) model—that encourage students to understand the important concepts, calculations, procedures, and safety considerations in the experiment through working together. (*Note:* Prelab workshops can be provided to students as more conventional prelab exercises if desired.) Further, by providing consistent and similar leading questions for each experiment, a pattern of thought will be developed and expected by the students as the semester progresses. Sample leading questions are shown below. A more detailed printable PLTL style template is available as an instructor's resource.

1. What are the goals of the lab? Summarize, in paragraph form, the procedures or sequence of events that will be followed to accomplish the goals. (What is the story of today's experiment and how will it be revealed?)

2. What calculations and equations will be needed? Define the terms in each equation and provide numerical values for constants and units for all constants and measurements.

3. Describe the appropriate safety procedures and why they are needed (be specific).

4. How will you dispose of chemical waste?

- Instructor's resources provided with each experiment outline variations that can be incorporated to enrich the student experience or tailor the lab to the resources/equipment available at different institutions. These resources can be found at www.mhhe.com/atomsfirstlab or www.mhhe.com/burdgeoverby.

- Each lab contains the following:

  1. The background section provides sufficient detail to understand the purpose of the experiment.

  2. A list of necessary materials enables planning the experiment.

  3. A procedures section—early labs have a detailed procedure to lead new students through the important steps, while procedures for selected later labs are structured more informally, requiring students to generate the details themselves.

  4. Data collection and analysis sections are structured to help students collect and manipulate the required data.

  5. Reflection questions at the end of each lab encourage students to interpret the results of the experiment and relate them to topics they have seen in their general chemistry lecture course.

  6. The connection question asks students to draw a connection between the lab and something they have experienced in everyday life or the world around them.

- Students are exposed to the essential techniques and equipment used in a chemistry laboratory setting. The authors have not included chapters on general laboratory safety guidelines or laboratory glassware and technique. It is assumed this important information will be provided by the instructor in a timely manner. We recommend a brief orientation be given prior to each lab in which technique and equipment can be demonstrated and safety discussed.

## ACKNOWLEDGMENTS

We would like to thank the following faculty members in the Chemistry Department at UTD for their numerous suggestions: Drs. Sandhya Gavva, Warren Goux, Steven Nielsen, Yanping Qin, Amandeep Sra, Claudia Taenzler, and Jie Zheng. We would also like to thank Anton Klimenko for his assistance in working out the details in the silver nanoprism lab.

Finally, we would like to acknowledge the enthusiasm and steady guidance of Shirley R. Oberbroeckling, the developmental editor for the first edition of our lab manual, who passed away this last year.

# 1

# Exploration of Matter Through Density Determinations: An Introduction to Basic Laboratory Measurements

## Objectives

- Become familiar with routine mass and volume measurements and their manipulations through calculation.
- Identify pure substances and mixtures using density determinations.

## Materials Needed

- Distilled water
- Volumetric pipette
- "Seawater"
- Unknown liquid
- Pennies (minted in 1983 or later)
- 50-mL graduated cylinder

## Safety Precautions

- Safety goggles/glasses must be worn at all times.

## Background

King Hiero II reigned over Syracuse, an ancient Greek city on Sicily, over 2300 years ago. To recognize the successes of the Syracusan army and to honor the gods, he provided a lump of gold to a goldsmith with the directive to prepare a golden crown. Upon receiving the crown, he became suspicious of its true composition despite the fact its weight was identical to that of the original mass of gold. In short, he was concerned that the goldsmith had replaced some of the gold in the crown with an equal mass of the considerably less expensive metal silver. To verify the composition of the crown, the king turned to the great Archimedes, a renowned mathematician and physicist (and possible relative of the king).[1-3] Because of the significance of the crown, Archimedes could not use techniques that would alter the crown in any way. Initially unsure of how to proceed, he found inspiration while climbing into a bathtub and noting the volume of water that his body displaced (Figure 1). It has been suggested that Archimedes shouted "Eureka" from his bathtub, setting forth the linkage of that word to discovery moments through current times. Based on this experience, Archimedes surmised that an object immersed in a liquid displaced a volume of liquid equal to the volume of the object. Thus, he obtained a sample of gold equal in mass to that of the crown, recognizing that two samples of pure gold of equal mass (the crown and his gold reference) will occupy the same volume, and explored their relative water

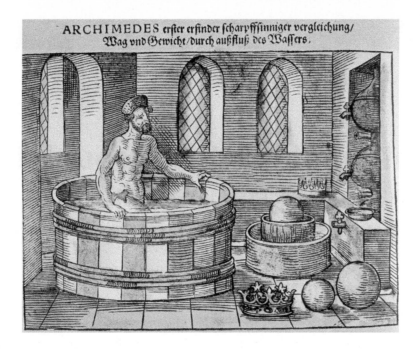

Figure 1. **Eureka!**

displacements. Through this simple, yet clever, approach, he determined that the crown was, indeed, not pure gold, and contained significant amounts of silver. The key to understanding the science behind Archimedes' experiment lies in the differing densities of gold and silver.

In this lab, you will explore basic laboratory procedures for acquiring mass and volume measurements with the subsequent calculation of density, the mass of a substance per unit volume, culminating in an Archimedes-inspired method for determining the composition of the modern penny. The density of an object can be calculated by dividing an object's mass by its volume:

$$D = \frac{\text{mass}}{\text{volume}}$$ **Equation 1**

The common unit for reporting the density of a solid or liquid is grams per milliliter, g mL$^{-1}$, or grams per cubic centimeter, g cm$^{-3}$, where 1 mL = 1 cm$^3$; gas densities are typically reported in g L$^{-1}$. A conceptual view of density is the amount of matter that occupies a specific area. Density is an *intensive* property, meaning it is characteristic of a pure substance and independent of the amount of substance present. As such, it can be used as a tool to identify a substance or the composition of a mixture of substances, as Archimedes showed. You will explore both of these applications through the identification of an unknown liquid and the inquiry into the composition of the modern penny. Please note that this laboratory involves acquiring measured quantities and using those quantities in calculations. As such, understanding the rules for recording the correct number of significant figures and the manipulation of significant figures through calculation are important!

## SIGNIFICANT FIGURES

For samples of pure water and aqueous solutions in laboratory glassware (graduated cylinders, burettes, pipettes, etc.), the volume is measured by noting the bottom of the liquid's meniscus, the curve in the surface of a confined standing body of liquid. In addition, the volume should be recorded to the correct precision of the volume-measuring device. This is typically done by recording all of the certain digits and estimating the last digit (see Figure 2). Taken together, all of the certain digits plus the estimated uncertain final digit describe the *significant figures* in the measurement. A "rule of thumb" is to record a measurement to 1/10 of the smallest calibration mark on the measuring device. Note that on a laboratory balance, the total significant figures, including the uncertain final digit, are displayed digitally.

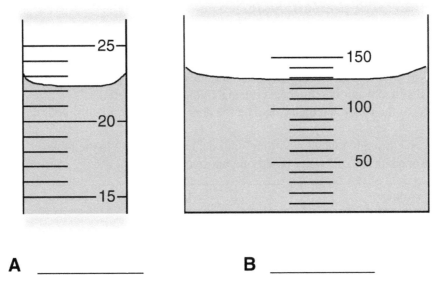

**A** _____       **B** _____

**Figure 2**   **The volumes in these two containers should be recorded to the precision of the volume measuring devices.**
Record the volumes in these two containers in the blanks under the drawings. Do your answers match ours (A: 22.3 mL; B: 129 mL)? Is 22.2 mL an acceptable answer for sample A? Yes, because the last digit is uncertain and, therefore, estimated based on the calibration marks of the glassware.

## SIGNIFICANT FIGURE RULES/CALCULATIONS

To describe the correct number of significant figures in a measurement, one should note the following:

- Nonzero integers always count.
- Leading zeros never count.
- Trailing zeros always count in the presence of a decimal point.

To an experimentalist, it is of utmost importance to preserve the precision of measured quantities in calculations by accurately keeping track of significant figures. The rules for significant figures in calculation are as follows:

1. For addition and subtraction operations, your answer should contain the same precision as the least precise measurement (i.e., the answer will have the same number of digits to the right of the decimal point as the original number with the smallest number of digits to the right of the decimal point).
2. For multiplication and division operations, your answer should contain the same number of significant figures as that of the measurement with the fewest total significant figures.

For example, in a representative density calculation, if a 2.320-g sample (four significant figures) is divided by 0.59 mL (two significant figures), the answer, 3.9 g mL$^{-1}$, must contain two significant figures. In contrast, if that same 2.320-g amount (last significant figure in the thousandths position) was added to a second mass of 20.2 g (last significant figure in the tenths position), the total mass would be 22.5 g, or the answer contains the same precision as the least precise measurement.

### References

1. Lewis AC. (1973). *Archimedes.* In *Encyclopedia of World Biography,* Vol. 1, pp. 219–223. McGraw-Hill, New York, NY.
2. (1914). *Vitruvius: The Ten Books on Architecture.* Trans. Morgan MH, pp. 253–254. Harvard University Press, Cambridge.
3. Baldwin J. (1905). *Thirty More Famous Stories Retold,* American Book Company, New York, NY. 235 pp.

## Procedure/Data Collection/Analysis

## I. BASIC LABORATORY OPERATIONS: DENSITY DETERMINATION OF WATER

1. Record the mass of your smallest clean, dry beaker in the data table below under Trial 1.
2. Pipette 5 mL of distilled water into the beaker using a volumetric pipette.
3. Measure the mass of the water and then calculate the density of water. Record your data under the Trial 1 heading.
4. Repeat procedures 2 through 4 while recording your data under Trial 2 in the table.
5. Determine the average density of water based on your two trials.

| Data Table 1: Water | Trial 1 | Trial 2 |
|---|---|---|
| Mass of beaker (g) | | |
| Tared mass of water (g) | | |
| Volume of water (mL) | | |
| Density of water (g mL$^{-1}$) | | |
| Average density of water (g mL$^{-1}$) | | |
| Laboratory temperature (°C) | | |

6. Given Equation 2 (below) and the accompanying table of densities (Table 1), calculate the % error in your experiment:

$$\% \text{ error} = \frac{|(\text{calculated density}) - (\text{known density})|}{(\text{known density})} \times 100 \qquad \textbf{Equation 2}$$

## Table 1 Density of Water at Various Temperatures.

| Water Density (g mL$^{-1}$) | Temperature (°C) |
|---|---|
| 0.998595 | 18 |
| 0.998405 | 19 |
| 0.998203 | 20 |
| 0.997992 | 21 |
| 0.997770 | 22 |
| 0.997538 | 23 |
| 0.997296 | 24 |
| 0.997044 | 25 |
| 0.996783 | 26 |
| 0.996512 | 27 |
| 0.996232 | 28 |

% error _____

7. OPTIONAL (consult your instructor): You may wish to add your average density of water to a class list and then calculate a class average density and class % error.

Class average density of water (g mL$^{-1}$) _____

% error _____

## II. BASIC LABORATORY OPERATIONS: DENSITY DETERMINATION OF SEAWATER

8. Perform steps 2 through 5 for the provided sample of seawater and record your data in the following table.

| Data Table 2: Seawater | Trial 1 | Trial 2 |
|---|---|---|
| Mass of beaker (g) | | |
| Tared mass of seawater (g) | | |
| Volume of seawater (mL) | | |
| Density of seawater (g mL$^{-1}$) | | |
| Average density of seawater (g mL$^{-1}$) | | |

## III. APPLICATION: DENSITY DETERMINATION OF AN UNKNOWN LIQUID

9. Perform steps 2 through 5 for a given unknown liquid and record your data in the following table.

| Data Table 3: Unknown | Trial 1 | Trial 2 |
|---|---|---|
| Mass of beaker (g) | | |
| Tared mass of unknown (g) | | |
| Volume of unknown (mL) | | |
| Density of unknown (g mL$^{-1}$) | | |
| Average density of unknown (g mL$^{-1}$) | | |

10. Using the density values for the set of common organic solvents shown in Table 2, identify your unknown liquid.

| Table 2    Densities of Common Organic Liquids. | |
|---|---|
| Substance | Density (g mL$^{-1}$ at 20 ºC) |
| Dichloromethane | 1.326 |
| Chloroform | 1.498 |
| Methanol | 0.791 |
| Hexane | 0.659 |
| Acetic acid | 1.049 |
| Ethyl acetate | 0.895 |
| Ethylene glycol | 1.115 |

Unknown _____

# IV. APPLICATION: DETERMINATION OF THE COMPOSITION OF A PENNY THROUGH A DENSITY STUDY

As shown in Table 3, the composition of the penny has varied since its introduction into U.S. currency in 1837. Your task is to determine the current percentages of copper and zinc that make up the modern penny (post-1982). You will do so by determining the density of a penny using a volume displacement method (see the density information in Table 4). Archimedes stated that an object immersed in a liquid displaces a volume of liquid equal to the volume of the object. If the object is more dense than water, then one can determine the volume of the object by sinking it in a known volume of water and recording the volume change.

| Table 3   Historical Data on the Composition of the Penny. |
| --- |
| 1837–1857: bronze (95% copper/5% tin/zinc) |
| 1857: 88% copper/12% nickel (coin was whitish) |
| 1864–1962: bronze (again) except for 1943 (zinc-coated steel) |
| 1962–1982: 95% copper/5% zinc (no tin) |
| 1982–present: ?% copper/?% zinc (YOU DETERMINE) |
| Source: www.usmint.gov |

| Table 4   Density and Cost of Representative Coinage Metals. | | |
| --- | --- | --- |
| | Density (g mL$^{-1}$) | Cost (U.S. $ per lb) |
| Copper | 8.96 | 3.07[1] |
| Nickel | 8.92 | 6.05[1] |
| Zinc | 7.13 | 0.83[1] |
| Silver | 10.5 | 279.42[2] |
| Gold | 19.3 | 18,014.06[3] |

[1]London Metal Exchange (July 8, 2013)
[2]Silver London PM Fix (July 8, 2013)
[3]Gold London PM Fix (July 8, 2013)

11. Obtain 25–35 pennies from the "pool" of pennies provided in the lab.
12. Fill a 50-mL graduated cylinder approximately half full with water and accurately record the initial volume of water in the following data table.
13. Record the exact number and mass of a sample of pennies numbering between 25 and 35.
14. Place the sample of pennies from step 11 into the 50-mL graduated cylinder one at a time and record the volume of the water in your data table. It is recommended that each penny be added to the graduated cylinder "edge first" so as not to trap air below the surface of the water.
15. Empty the graduated cylinder, dry the pennies thoroughly, and repeat the experiment a second time using a different set of pennies.
16. Calculate the average density of a penny.

| Data Table 4: The Penny | Trial 1 | Trial 2 |
|---|---|---|
| Number of pennies | | |
| Total mass of pennies (g) | | |
| Initial volume of water (mL) | | |
| Final volume of water (mL) | | |
| Volume the pennies occupy (mL) | | |
| Density of a penny (g mL$^{-1}$) | | |
| Average density of a penny (g mL$^{-1}$) | | |

17. The percent composition of the penny can now be determined by noting that the average density determined in step 16 represents a weighted average of densities of the two component metals, copper and zinc. This weighted average can be expressed mathematically as follows:

$$x \text{ (copper density)} + (1 - x) \text{ (zinc density)} = \text{average density of the penny} \qquad \textbf{Equation 3}$$

where

$$x = \text{the fraction of the penny that is comprised of copper}$$
$$1 - x = \text{the fraction of the penny that is comprised of zinc}$$

Using the densities from Table 4 and Equation 3 determine the values of $x$ and $1 - x$ and then convert the decimal values to percentages by multiplying each by 100. Report your final answer below.

Composition of the modern penny = _____ % zinc and _____ % copper

## Reflection Questions

1. As shown in Table 1, the density of water is temperature dependent. Why?

2. Considering your results in Parts I and II, is it easier for you to float in the ocean or in a freshwater lake? Explain.

3. If some of the unknown liquid evaporated in Part III after you added it to your beaker, but before you recorded its mass, what effect would this have on your reported density?

4. Could you use the procedures and calculations in Part IV to identify the % composition of a nickel coin (75% Cu/25% Ni)? Why or why not?

5. Copper metal and zinc metal react differently in the presence of certain acids. For example, while both react with concentrated nitric acid and dissolve completely, only zinc reacts and dissolves in dilute hydrochloric acid. Given this information, how else could you accurately determine the composition of a modern (post-1982) penny?

   a. What volume of liquid, expressed in $cm^3$, would a 16-lb bowling ball displace (diameter = 8.5 in). The following information may be useful for you:

   $$1 \text{ in} = 2.54 \text{ cm}, \ 16 \text{ oz} = 1 \text{ lb}, 28.3 \text{ g} = 1 \text{ oz, volume of a sphere} = \left(\frac{4}{3}\right)\pi r^3$$

b. Considering your experiences *handling* a bowling ball, what obvious error is in the answer in 5a?

c. Would a ball equal in volume to the bowling ball described in part a, but weighing only 1 lb, displace the same volume of water as the bowling ball?

6. Considering the data in Table 4, would equal masses of gold and copper result in the same volume change if added to water?

7. Give two reasons why the federal government might have changed the composition of the penny in 1982. (Hint: See Tables 3 and 4.)

## Connections

Based on your experience in this lab, draw a connection to something in your everyday life or the world around you (something not mentioned in the background section):

# 2

# The Discovery of Chemical Change Through the Chemistry of Copper: An Observational Preview of First-Semester General Chemistry

## Objectives

- Observe chemical change through a series of chemical reactions.
- Learn simple mass and volume measuring techniques.
- Understand and use the law of conservation of mass.
- Learn the proper techniques for the handling and disposal of chemicals.

## Materials Needed

- Copper wire
- Zinc metal (pieces)
- Concentrated nitric acid ($HNO_3$)
- 3 $M$ sulfuric acid ($H_2SO_4$)
- 6 $M$ sodium hydroxide (NaOH)
- 6 $M$ hydrochloric acid (HCl)
- Acetone
- Hot plate
- Water
- 250-mL beaker
- 50-mL graduated cylinder
- 10-mL graduated cylinder
- Laboratory balance

## Safety Precautions

*Note: This is an introductory lab involving chemical reactions. You will be working with strong acids, bases, oxidizing agents, and flammable substances. While general safety information is provided below, techniques for safe handling of these substances and their disposal will be described by the instructor.*

- Gloves and safety goggles/glasses should be worn during this lab.
- Concentrated nitric acid is both a strong acid and strong oxidizing agent. It must be stored and used in a laboratory hood. Avoid contact as nitric acid can cause severe burns and a yellowing of the skin (due to a xanthoproteic reaction involving nitric acid and proteins in skin cells).

- The reaction of copper metal with nitric acid in step 3 of this lab is a vigorous reaction that produces nitrogen dioxide, a red-brown toxic gas. This reaction must be done entirely within a laboratory hood.
- Sodium hydroxide is a strong base that can cause burns. Avoid contact with body tissues.
- Hydrochloric acid is toxic by ingestion and inhalation and corrosive to skin and eyes; avoid contact with body tissues.
- In step 9 of this lab, hydrogen gas will be generated; as hydrogen is flammable, keep all heat and flames away from your reaction vessel.
- Acetone is flammable. Keep away from open flames and heat sources.
- Dispose of waste materials as indicated by your instructor.

## Background

Understanding in science is tied to observation, hypothesis, experimentation, and refinement of the hypothesis until data-driven conclusions are found that describe natural phenomena. The key here is the recognition that data, born from experiment and observation, give rise to both a narrowly focused (e.g., the development of a new anticancer drug) and broader (theories) understanding of the world around us. This lab includes many of the concepts and skills that will be introduced in subsequent chapters and, thus, can be reconsidered throughout the semester for increased understanding of the chemical principles involved in its procedures and consequent observations. In its simplest form, this lab is a discovery experience in which chemical change is in full display, as shown through color and physical state changes. In addition to an overarching goal of understanding the law of conservation of mass through experiment, it is hoped that you will enjoy the opportunity to observe and control chemical change.

The law of conservation of mass is fundamental to science and states that matter cannot be created or destroyed. Its application to chemical reactions is quite profound in that all matter (every atom) at the beginning of a reaction must be exactly accounted for after the reaction is complete. So, for example, the reaction of one carbon atom with two oxygen atoms will produce exactly one molecule of carbon dioxide, $CO_2$. This statement can be summarized in a *balanced* chemical equation as noted below:

$$C + 2O \rightarrow CO_2$$

The term "balanced" reflects that the same number of atoms of each type (C and O in this case) are present on both sides of the equation. The law of conservation of mass must be obeyed as the chemical change involving the combination of the elements carbon and oxygen to produce carbon dioxide takes place. In other words, the total mass of the carbon and oxygen present at the beginning of the reaction must equal the total mass of carbon dioxide produced.

While the notion that matter is conserved or that it cannot be created or destroyed dates from ancient Greek times, the law of conservation of mass was not formally stated until the late eighteenth century by Antoine Lavoisier based on his studies involving the reactions of metals with oxygen gas. Lavoisier (Figure 1) showed that the mass of oxygen and a metal sample, tin, was equal to the mass of metal oxide product formed upon their reaction:

$$2Sn + O_2 \rightarrow 2SnO$$

metal (tin)    air    metal oxide

Because oxygen is a gas, Lavoisier needed to perform these experiments in sealed vessels to precisely measure masses. As a historical note, these experiments helped disprove the existence of phlogiston, a theorized substance thought to be a part of matter that was lost as it burned. Prior to Lavoisier's seminal experiments, scientists were puzzled by the decreasing mass of, for example, a piece of wood as it burned or reacted with oxygen, attributing this mass loss to "phlogiston," which was thought to be released into the atmosphere as the substance burned. Equally confusing was the observation that some

**Figure 1** **Antoine Lavoisier, the father of modern chemistry.**

substances, such as metals, appeared to gain mass as they reacted with the air (oxygen, specifically), suggesting that the released phlogiston in these metals must have a negative mass! Through careful experiment, Lavoisier demonstrated that, in fact, no mass was lost or gained during a chemical reaction and phlogiston was fantasy.

Antoine Lavoisier (1743–1794) is generally considered the father of modern chemistry. Prior to his work, chemistry was more of a qualitative science, full of observation but lighter in verifiable explanation. Lavoisier's precise measurements and ability to provide clarity through analysis of his own data and, in particular, from the works of others, allowed for hypotheses to be tested experimentally with the eventual development of increased understanding and theories. Thus, he is credited with advancing chemistry as a *quantitative* science. Without quantitation, the scientific method, rooted in reproducible experimentation, could not have developed into the standard used by scientists to explain the world and make remarkable discoveries. Lavoisier was a passionate scientist with broad interests. He made fundamental contributions to the following: an understanding of reactions involving oxygen gas, including combustion reactions and respiration; nomenclature (or the naming of chemical compounds); the atomic composition of water (including naming the elements hydrogen and oxygen); the definition and classification of substances as elements; allotropes, or different forms of the same element by identification of diamond as a pure carbon substance; the relationship between energy and chemical reactions; and the construction of the metric system. He is even credited with authoring the first textbook on chemistry. As a French aristocrat with a vested interest in an institution responsible for collecting taxes (the unpopular Ferme Générale), Lavoisier was ultimately beheaded as a traitor by the revolutionaries during the French Revolution.

## THIS LAB

In this lab, you will attempt to verify the law of conservation of mass by converting a sample of copper metal into various compounds through a series of chemical reactions, ultimately returning it to its metallic copper form (Figure 2). If your laboratory technique is sound and you are a keen observer, then you should be able to demonstrate that, while the copper will have changed forms several times, the mass that you produce at the end should be similar (identical according to Lavoisier and the law of conservation of mass) to the mass with which you began. Copper, like metals in general, is a good electrical and thermal conductor that can be drawn into wires (ductile) and flattened into sheets (malleable). A softer metal, copper's hardness can be improved through alloys (mixtures with other metals) involving zinc or tin to make

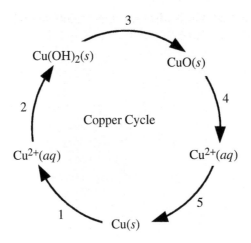

**Figure 2 The copper cycle.**
Five sequential reactions (numbered 1 through 5) will be used to convert copper metal into various forms, with the final step regenerating copper in its metallic form.

brass and bronze, respectively. Copper is an essential element for plants and animals, though not in metallic form but as the copper(II) ion. Copper(II) salts contain the copper atom in an ionized or charged form and are typically blue to green in color.

In a chemical reaction, substances interact with each other to produce new substances with properties distinct from the original ones. In other words, the observation of chemical change defines a chemical reaction. You will explore six distinct chemical reactions; each one is described below. The details for the complete understanding of these reactions and the substances that are present in them will unfold over the course of this semester. However, by introducing some of the language and concepts of chemical reactions here, a seed is planted for you to begin focusing on the need to sequentially build up an understanding of atomic structure, bonding, physical and chemical properties, and chemical reactivity—topics covered extensively in first-semester general chemistry. As you will soon learn, the properties and reactivity of all matter around and in you are directly linked to their structures.

*Reaction 1*: This reaction is an example of an oxidation-reduction reaction in which the copper metal, containing neutral copper atoms, is oxidized (loses electrons) to produce the blue $Cu^{2+}$ ion in solution (see Figure 3). Oxidation-reduction reactions or "redox" reactions are an important and common class of reactions in which electrons are transferred between reacting species.

 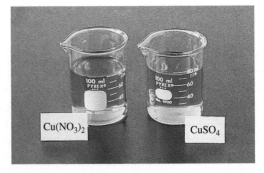

**Figure 3 Copper metal (left) and aqueous solutions of blue $Cu^{2+}$ ions (right).**

*Reaction 2*: The $Cu^{2+}$ ion is combined with hydroxide, $OH^-$, a polyatomic ion, to form an insoluble solid or precipitate. The product is named copper(II) hydroxide, $Cu(OH)_2$. This is an example of another common type of reaction, typically involving ions, called a precipitation reaction.

*Reaction 3*: In this reaction, the solid product from reaction 2 loses water through heat to give a new $Cu^{2+}$-containing compound, copper(II) oxide or CuO.

*Reaction 4*: The solid product from reaction 3 is digested by sulfuric acid in an acid-base or proton-transfer reaction, a very common type of reaction that involves compounds classified as acids and bases. The result is the presence of the $Cu^{2+}$ ion in solution as produced by reaction 1.

*Reaction 5*: This redox reaction regenerates copper in its original metallic form. A different metal, zinc, is used as the source of electrons to reduce (add electrons to) the $Cu^{2+}$ ions in solution and cause the formation of the insoluble copper metal. As zinc is the source of electrons, it becomes oxidized and goes into solution as the colorless $Zn^{2+}$ ion. Important to the integrity of the final product, the reverse reaction involving the newly formed $Zn^{2+}$ ion and copper metal does not occur. This demonstrates that the two metals copper and zinc have distinct reactivities, with zinc being the more active metal.

*Reaction 6*: This final reaction does not involve copper at all, but allows for the excess zinc metal used in reaction 5 to be consumed so that it will not affect the final mass of the copper product. The reaction involves the oxidation (a redox reaction, again) of the zinc metal with hydrochloric acid. Hydrochloric acid selectively reacts with the more active zinc metal and not the copper product, again demonstrating that different substances (metals, in this case) have different chemical properties.

## References

1. Bryson, B. (2005). *A Short History of Nearly Everything*. Broadway Books, New York. pp. 125–128.
2. Donovan, A. (1993). *Antoine Lavoisier: Science, Administration and Revolution*. Blackwell Publishers, Oxford.

## Procedure

1. Record the mass of a 250-mL beaker.
2. Add approximately 0.5 g of copper wire to the beaker and record the combined mass of the beaker plus the copper metal. Place the beaker containing the copper sample onto a hotplate in a laboratory hood.
3. In the hood, carefully add 2.5 mL of concentrated nitric acid to the copper sample and cover the beaker with a watch glass. Gently heat the mixture using a hot plate until it becomes viscous (viscosity is a measure of a liquid's resistance to flow; a viscous liquid resists flow and can be characterized more casually as appearing thick). Record all observations.
4. After allowing the reaction mixture to cool, add 5 mL of deionized water and stir. Then, slowly add 5 mL of 6 *M* NaOH with continued stirring. Record all observations.
5. Add 25 mL of deionized water to the beaker and heat to boiling for several minutes. Be careful to avoid splattering of the beaker contents during the boiling process. Placement of a stirring rod in the beaker during the boiling process will reduce the potential for splattering. Heat for an additional 5 to 10 minutes after any changes in the reaction mixture appear complete. Remove the beaker from the heat and record any observations.
6. Decant the liquid into a beaker without losing any of the solid. Add 15 mL of deionized water to the solid residue making sure to wash the inside walls of the container. Decant the water and repeat the washing of the solid with a second 15-mL portion of water. Dispose of the decanted washings into the appropriate waste container as noted by your instructor. The remaining washed solid is to be used in the next step.

7. Add 15 mL of 3 $M$ $H_2SO_4$ to the solid from step 6 and gently heat until all of the solid has dissolved. Record any observations.
8. Add zinc metal in small pieces to the solution formed in step 7, not exceeding a total mass of 1.5 g of zinc. Add pieces of zinc as necessary until the solution is colorless. You may need to hold a piece of white paper behind the beaker as you attempt to determine if it is truly colorless. If the solution remains colored, you must add additional zinc until a colorless solution results. Record any observations.
9. Decant the solution, being careful to leave behind the solid residue. To this solid, add 5 mL of 6 $M$ HCl, which will react with excess zinc from step 8, and record any observations. Decant the solution, being careful to leave behind your final product, pure copper. The final copper product should then be washed with two 25-mL portions of water. All water washes should be disposed of in a waste container as noted by your instructor.
10. To aid the drying of the final copper product, wash it with one or two separate 10-mL portions of acetone, a low-boiling organic solvent that will remove any remaining water. Be sure to rinse the sides of the container with the acetone washes. The acetone washings (decanted liquid) should then be discarded in a waste container labeled for organic waste.
11. Label your sample by writing the sample information on a small piece of paper and placing the paper in the beaker.
12. Place your labeled sample in a designated area for drying until the next lab period.
13. At the beginning of the next lab period, remove your paper label and record the mass of the beaker and its contents. Describe the appearance of the contents in your beaker and have your instructor initial your description.

## Data Collection

1. Mass of 250-mL beaker, step 1: _____

2. Mass of 250-mL beaker plus the initial copper sample, step 2: _____

3. Observations for reaction 1, step 3:    $Cu(s) + H^+(aq) + NO_3^-(aq)$

4. Observations for reaction 2, step 4:    $Cu^{2+}(aq) + OH^-(aq)$

5. Observations for reaction 3, step 5:    $Cu(OH)_2(s) + heat$

**6.** Observations for reaction 4, step 7:     $CuO(s) + H^+(aq)$

**7.** Observations for reaction 5, step 8:     $Cu^{2+}(aq) + Zn(s)$

**8.** Observations for reaction 6, step 9:     $Zn(s) + H^+(aq)$

**9.** Mass of 250-mL beaker plus the final copper sample, step 13:     _____

**10.** Description of beaker contents, step 13:

_____

instructor's initials

## Analysis

**11.** Mass of initial copper sample:     $mass_{copper} = mass_{beaker + copper} - mass_{beaker}$

$$= (\#2) - (\#1) \qquad = \underline{\hspace{2cm}}$$

**12.** Mass of final copper sample:     $mass_{copper} = mass_{beaker + copper} - mass_{beaker}$

$$= (\#10) - (\#1) \qquad = \underline{\hspace{2cm}}$$

**13.** $\%error = \dfrac{|(\text{mass of copper})_{initial} - (\text{mass of copper})_{final}|}{(\text{mass of copper})_{initial}} \times 100$

$$= \dfrac{|(\#11) - (\#12)|}{(\#11)} \times 100 \qquad\qquad = \underline{\hspace{2cm}}$$

1.  A balanced chemical equation describes a chemical reaction consistent with the law of conservation of mass. To balance equations, one needs to understand that a chemical formula contains the component elements of the substance in the ratio in which they are present. For example, the formula for water is $H_2O$, meaning that two H atoms and one O atom comprise a single water molecule. A balanced chemical equation must demonstrate that the same number of atoms of each type are present before and after the reaction takes place. In other words, no atoms are created or destroyed during the reaction; they are simply rearranged to produce new substances. Coefficients are used in front of the formulas as necessary to give rise to a balanced equation. For example, as shown below, hydrogen and oxygen can be combined to produce water, a reaction of considerable interest as an alternative to our current energy sources because of the amount of energy that is released into the environment over the course of the chemical change. The equation is shown in unbalanced and balanced forms. Note the use of coefficients to multiply selectively the formulas of the substances to reflect the law of conservation of mass. The formulas, however, must not be changed during the balancing process! Finally, a balanced equation typically contains coefficients in the smallest possible whole integer values.

$$H_2(g) + O_2(g) \rightarrow H_2O(g) \text{ (unbalanced)}$$

$$2H_2(g) + O_2(g) \rightarrow 2H_2O(g) \text{ (balanced)}$$

The following six chemical equations describe the six reactions that you used to convert copper into various compounds and ultimately reproduce the original copper sample. Consistent with the law of conservation of mass, determine if each of the given equations are balanced. Balance those that are unbalanced by including the appropriate coefficients.

Reaction 1:     $Cu(s) + 4H^+(aq) + 2NO_3^-(aq) \rightarrow Cu^{2+}(aq) + 2NO_2(g) + H_2O(l)$

Reaction 2:     $Cu^{2+}(aq) + OH^-(aq) \rightarrow Cu(OH)_2(s)$

Reaction 3:     $Cu(OH)_2(s) \rightarrow CuO(s) + 2H_2O(l)$

Reaction 4:     $CuO(s) + H^+(aq) + HSO_4^-(aq) \rightarrow Cu^{2+}(aq) + SO_4^{2-}(aq) + H_2O(l)$

Reaction 5:     $Cu^{2+}(aq) + Zn(s) \rightarrow Cu(s) + Zn^{2+}(aq)$

Reaction 6:     $Zn(s) + H^+(aq) \rightarrow Zn^{2+}(aq) + H_2(g)$

2. The burning of wood in a campfire may be viewed as violating the law of conservation of mass. The burning of wood is a combustion reaction (a reaction with oxygen) in which the wood material is combined with the oxygen in the air to produce carbon dioxide gas and water vapor (shown symbolically below). Consider what you observe as the wood burns. Over time, the mass of solid material (wood) in the fire pit decreases. This puzzled scientists prior to Lavoisier.

$$\text{wood} + \text{air (oxygen)} \rightarrow CO_2(g) + H_2O(g)$$

Suggest a set of experimental conditions that you would use to demonstrate that the burning of wood does, in fact, obey the law of conservation of mass.

3. How would the following errors affect the final mass of copper in this experiment?
   (a) Small amounts of water remained in the beaker after washing with acetone and drying your sample for a week (at step 13).

   (b) While decanting in step 6, a small amount of solid is lost with the decanted liquid.

   (c) You neglected to perform reaction 6 (step 9).

4. Acid is needed in reaction 6 to consume excess zinc metal. Would nitric acid (used in reaction 1) be an acceptable replacement for the hydrochloric acid in reaction 6? Why or why not?

5. Consider the reaction described by the balanced equation below. If the combined mass of $CuSO_4$ and $Zn(s)$ initially equals 3.84 g, what would the combined mass of the products, Cu and $ZnSO_4$, be assuming the reaction goes to completion with only products present? Do you have enough information to answer the question?

$$CuSO_4 + Zn \rightarrow Cu + ZnSO_4$$

## Connection

Based on your experience in this lab, draw a connection to something in your everyday life or the world around you (something not mentioned in the background section):

# 3

# Light and Nanotechnology: How Do We "See" Something Too Small to See?

## Objectives

- Measure the width of a human hair using a laser pointer and diffraction.
- Synthesize silver nanoparticles, and investigate how color is related to particle size.

## Materials Needed

- Red laser pointer ($\lambda = 633$ nm)
- Tape measure
- Scotch tape
- Somebody with hair!
- High-purity distilled water (18.2 $\Omega M$ cm resistivity works well)
- Stock solutions (to be prepared by instructor)
  — $1.25 \times 10^{-2} M$ sodium citrate
  — $3.75 \times 10^{-4} M$ silver nitrate ($AgNO_3$)
  — $5.68 \times 10^{-2} M$ hydrogen peroxide ($H_2O_2$; should be prepared fresh each day)
  — $1.85 \times 10^{-5} M$ potassium bromide (KBr)
  — $5.0 \times 10^{-3} M$ sodium borohydride ($NaBH_4$; should be prepared fresh each day)
- Four large (25-mL) test tubes
- Stoppers or parafilm to seal test tubes
- Method for accurately measuring the absorbance of samples. Possible options include:
  — Vernier LabPro handheld device with a SpectroVis Plus spectrophotometer attachment. The associated software LoggerPro must be installed on a computer to allow downloading and manipulation of data. Information about the Vernier device and associated peripherals and software can be obtained from the website http://www.vernier.com (last accessed July, 2013).
  — UV/Vis spectrophotometer
  — Spectronic 20 spectrophotometer
- Cuvettes compatible with the device used to measure absorbance

## Safety Precautions

- Gloves should be worn during this lab.
- $AgNO_3$ is corrosive and causes burns if it comes in contact with the skin and eyes.
- $NaBH_4$ is flammable and toxic; when wet it may release hydrogen gas, so **do not** keep $NaBH_4$ solutions in tightly sealed vessels to avoid explosion of the container.

- Sodium citrate may act as an irritant to the skin, eyes, and respiratory tract.
- Hydrogen peroxide is corrosive and causes burns to the eyes, skin, and respiratory tract.
- Silver nanoparticle solutions should not be kept tightly capped due to possible presence of unreacted borohydride.
- Do not shine a laser pointer at anyone's eyes.

## Background

### MATERIALS AT THE NANOSCALE

Nanoscience . . . nanotechnology. Nearly everyone has heard something in the news about the nanoscale world promising the next big scientific revolution, from the development of new cures for diseases to solving food and energy problems. It can be difficult these days to separate the science from the hype and the truth from the misinformation. Actually, several of the promises of nanotechnology are already being realized.

In fact, nanomaterials are finding ever-widening use in our everyday lives, ironically as they have for thousands of years. In ancient times, metal nanoparticles were used as pigments; church windows were colored yellow using silver nanoparticles, and red was generated using gold nanoparticles. Today nanoparticles are found in a host of products, including cosmetics, skin creams and sunblock, and stain-resistant fabrics. Carbon nanotubes—small cylinders formed from carbon—are being added to bicycle frames, tennis rackets, baseball bats, golf clubs, and bowling ball coatings to make them lighter and stronger, as well as to microelectronics due to their superior conduction properties. Nanoparticles are also finding their way into biomedical applications. Colloidal silver has supposed antibacterial activity,[1-3] which was recognized even in ancient Greek and Babylonian cultures. In the early 1900s in American hospitals, silver nitrate was used in the eyes of newborn babies to protect against blindness. And NASA used a silver-based water purification system in its space shuttles.

Nanoscience is the study of objects and processes that take place at the nanometer scale. One nanometer is equivalent to 1 *billionth* of a meter. Objects that contain at least one dimension that is below 100 nm are considered nanoscale objects. In this size regime, we are typically dealing with relatively small collections of atoms or molecules. The reason for the interest in the nanoscale stems from the fact that the properties of matter can differ significantly from those observed for bulk materials. The understanding of these special properties at the nanoscale depends on our ability to characterize particles that are too small to see or easily manipulate. A main tool for probing nanoparticle structure and properties is light. The goal of this lab is to utilize various properties of light to study two types of very small objects—silver nanoparticles that are < 100 nm in diameter and single strands of human hair (not a nanoscale object, yet too small to measure using standard measurement tools with which you are familiar). A brief description of the properties of light is provided next.

### PROPERTIES OF LIGHT

Light can be described as being produced by the oscillating motion of an electric charge (as was done by Maxwell in 1873). This generates oscillating magnetic and electric fields perpendicular to each other that propagate through space as waves—these waves are called **electromagnetic (EM) radiation** (Figure 1). In general, waves are characterized by three basic features (Figure 2): (1) *wavelength* ($\lambda$), which is the distance between two identical points in the wave (e.g., peak to peak distance); units are usually meters (m) or nanometers (nm); (2) *amplitude,* or the distance from a midline of the wave to any point on the wave; and (3) *frequency* ($v$), which is the number of regularly occurring events (e.g., wavelengths) that occur per unit of

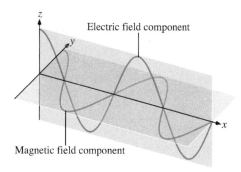

**Figure 1  Illustration showing electromagnetic wave.**

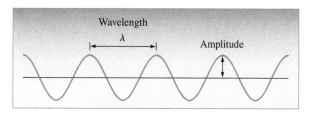

**Figure 2  Wave with wavelength and amplitude labeled.**

time; units are usually inverse seconds ($s^{-1}$ = hertz = Hz). $\lambda$ and $v$ can be related to the speed of the EM wave using Equation 1:

$$c = \lambda v \qquad \qquad \textbf{Equation 1}$$

where $c$ = speed of light = $2.998 \times 10^8$ m s$^{-1}$ (in a vacuum). We can also relate $\lambda$ and $v$ to the energy associated with a given wave:

$$\Delta E = hv \qquad \qquad \textbf{Equation 2}$$

$$\Delta E = (hv)/\lambda \qquad \qquad \textbf{Equation 3}$$

where $h$ = Planck's constant = $6.63 \times 10^{-34}$ J s.

EM radiation is a continuous spectrum of wavelengths (frequencies) with all wavelengths present (no gaps). Between ~400 and 800 nm is the region we call visible light (can be detected by the human eye), with red at the long wavelength (low frequency, low energy) and violet at the short wavelength (high frequency, high energy) (Table 1 and Figure 3).

White light (e.g., sunlight) contains a continuous spectrum of all visible wavelengths and is called *polychromatic* light. When light hits an object, different frequencies can be selectively absorbed (gathered or taken up), reflected (bounced back), or transmitted (allowed to pass through), depending on the frequencies of light, the chemical composition of the object, and even the object's size. The eye interprets the color of the object based on what wavelengths are reflected or transmitted. If none of the light is absorbed, then the object appears white since all visible wavelengths are reflected/transmitted. If all wavelengths are absorbed, then the object appears black (no wavelengths reflected/transmitted). However, if the object absorbs only one wavelength, such as ~550 nm (green

| Table 1 Wavelengths Associated with Various Colors of Visible Light. | |
|---|---|
| Color | Wavelength (nm) |
| Violet | ~400 to 450 |
| Blue | ~450 to 480 |
| Cyan | ~480 to 520 |
| Green | ~520 to 560 |
| Yellow | ~560 to 600 |
| Orange | ~600 to 640 |
| Red | ~640 to 700 |

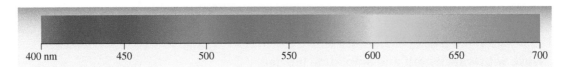

Figure 3   **Visible light spectrum.**

light), then the object will appear red. Each color has a *complementary color*—this is best shown in a color wheel, where the complement of each color is directly across the circle (Figure 4).

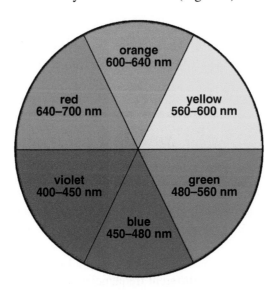

Figure 4   **Color wheel.**

When a specific color of light is absorbed, all other wavelengths are reflected/transmitted, and the complementary color is the one that the eye perceives (thus, orange is seen when blue light is absorbed). Other scenarios can give rise to specific colors; for example, an object might reflect green light and absorb all other light (object would appear green), or an object might transmit blue light and absorb all others (object would appear black on one side, and blue on the other).

## SEPARATING LIGHT

There are several methods that can be used to take polychromatic light and separate it into individual colors (i.e., a spectrum). The best-known method involves a prism, which utilizes refraction to bend different wavelengths of light to different degrees; the different paths taken by each wavelength lead to color separation. The same effect is behind the formation of a rainbow when sunlight passes through moisture in the air.

Another method utilizes a diffraction grating—a large number (e.g., 1000 grooves per mm) of tiny parallel grooves (or lines) that bend light waves, sending them in different directions based on wavelength. White light passing through a diffraction grating will generate a rainbow, with more grooves generating a better separation of the colors." [Note: in the lab "Shedding Light on the Structure of the Atom" we will be utilizing a diffraction grating to separate light from various sources.]

In single-slit diffraction, light passes through a narrow slit of width $w$ and strikes a screen at a distance $L$ from the slit (Figure 5).

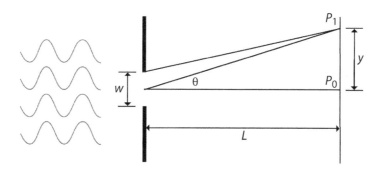

**Figure 5   Diffraction of light through a slit.**

The illustration shows coherent light (all waves with same $\lambda$) such as laser light from a laser pointer. The light waves bend around the slit edges and spread. Where two or more waves arrive in phase at the same point at the same time, the waves add together (constructive interference) and a bright spot is generated. In contrast, destructive interference occurs when two or more waves arrive at a point out of phase and subtract from each other, generating a darker region. The result is a pattern of bright and dark regions. In Figure 5, $P_0$ represents the point of maximum intensity where the waves pass straight through the slit, while point $P_1$ represents a dark point of complete destructive interference.

Equations 4 and 5 describe the relationship between $\lambda$, $L$, $y$, and $w$:

$$w\sin(\theta_n) = n\lambda \qquad\qquad \textbf{Equation 4}$$

$$\tan(\theta_n) = y_n/L \qquad\qquad \textbf{Equation 5}$$

where $n = 1, 2, 3, \ldots, y_n =$ distance from $P_0$ to $P_n$, and $\theta_n$ is the angle corresponding to $P_n$. So, if we measure the distances $L$ and $y_n$, and if we know $\lambda$, then we can calculate the width of the slit ($w$). In this lab, a variation of single-slit diffraction is used to measure the width of a human hair; instead of a single slit in a wall, we will be using a single human hair of width $w$ in open air to generate a similar diffraction pattern.

# SILVER NANOPARTICLES

Silver nanoparticles (as well as nanoparticles of gold and copper) have colors that depend on their size and shape.[4–9] The color is generated by **surface plasmon resonance.** Light that strikes the particles is absorbed and creates oscillations in the freely moving (conduction) electrons on their surfaces. The specific wavelength(s) that is(are) absorbed depend(s) on the size and shape of the particles.

In this lab, you will be synthesizing prism-shaped silver nanoparticles. In general, this process involves the conversion of $Ag^+$ ions to $Ag(s)$ through a *reduction reaction* (you will learn more about reduction reactions in a later lab; for now, suffice it to say that a reduction involves adding one or more electrons to an atom or ion). For this process, sodium borohydride ($NaBH_4$) is used as the reducing agent:[4,7,10–13]

$$8AgNO_3(aq) + NaBH_4(aq) + 4H_2O(l) \rightarrow 8Ag(s) + Na[B(OH)_4](aq) + 8HNO_3(aq)$$

The $Ag(s)$ atoms will stick together to form small particles. This process will continue, with the particles getting larger, until the particle surface is covered or coated with a charged species that limits further addition of Ag atoms. The added citrate acts to buffer or control the pH of the solution and, by binding to the surface of the growing nanoparticles, to limit their growth. The hydrogen peroxide acts as an "etching agent," removing silver atoms from less stable nanoparticles in a process called an *oxidation reaction* (involves removing one or more electrons from an atom or ion; it is the opposite reaction as a reduction).

$$2Ag(s) + H_2O_2(aq) + 2H^+(aq) \rightarrow 2Ag^+(aq) + 2H_2O(l)$$

The borohydride and $H_2O_2$ work in concert under the conditions used here, resulting in the formation of silver nanoprisms.[12–14]

Bromide is added to control the size of the nanoprisms.[11,14,15] As more bromide is added, the size of the nanoprisms decreases, as can be seen in Table 2.

## Table 2   Effect of Bromide on Nanoprism Size.[11]

| Volume 18.5 $\mu$M KBr Solution Added (mL) | Expected Color | Average Width (lateral dimension; nm) |
|---|---|---|
| 1.0 | Blue | ~64 |
| 1.5 | Orange-red | ~35 |
| 2.0 | Yellow | ~20 (spherical) |

## *References*

1. Shahverdi AR, Fakhimi A, Shahverdi HR, Minaian MS. Synthesis and Effect of Silver Nanoparticles on the Antibacterial Activity of Different Antibiotics against *Staphylococcus aureus* and *Escherichia coli*. *Nanomedicine.* 2007; 3: 168–171.
2. Pal S, Tak YK, Song JM. Does the Antibacterial Activity of Silver Nanoparticles Depend on the Shape of the Nanoparticle? A Study of the Gram-Negative Bacterium *Escherichia coli. J. Applied and Environmental Microbiology.* 2007; 73: 1712–1720.
3. Guzmán MG, Dille J, Godet S. Synthesis of Silver Nanoparticles by Chemical Reduction Method and Their Antibacterial Activity. *Int. J. Chem. Biomol. Engineering.* 2009; 2: 104-111.
4. Solomon SD, Bahadory M, Jeyarajasingam AV, Rutkowsky SA, Boritz C. Synthesis and Study of Silver Nanoparticles. *J. Chem. Ed.* 2007; 84: 322–325.
5. Wang G, Shi C, Zhao N, Du X. Synthesis and Characterization of Ag Nanoparticles Assembled in Ordered Array Pores of Porous Anodic Alumina by Chemical Deposition. *Mater. Lett.* 2007; 61: 3795–3797.
6. Campbell DJ, Xia Y. Plasmons: Why Should We Care? *J. Chem. Educ.* 2007; 84: 91–96.
7. Liz-Marzan LM. Nanometals: Formation and Color. Mater. *Today.* 2004; 7: 26–31.

8. Sun Y, Xia Y. Gold and Silver Nanoparticles: A Class of Chromophores with Colors Tunable in the Range from 400 to 750 nm. *Analyst.* 2003; 128: 686–691.

9. Mock JJ, Barbic M, Smith DR, Schultz DA, Schultz S. Shape Effects in Plasmon Resonance of Individual Colloidal Silver Nanoparticles. *J. Chem. Phys.* 2002; 116: 6755–6759.

10. Fang Y. Title: Optical Absorption of Nanoscale Colloidal Silver: Aggregate Band and Adsorbate-Silver Surface Band. *J. Chem. Phys.* 1998; 108: 4315–4318.

11. Frank AJ, Cathcart N, Maly KE, Kitaev V. Synthesis of Silver Nanoprisms with Variable Size and Investigation of Their Optical Properties: A First-Year Undergraduate Experiment Exploring Plasmonic Nanoparticles. *J. Chem. Ed.* 2010; 10: 1098–1101.

12. Millstone JE, Hurst SJ, Metraux GS, Cutler JI, Mirkin CA. Colloidal Gold and Silver Triangular Nanoprisms. *Small.* 2009; 5: 646–664.

13. Metraux GS, Mirkin CA. Rapid Thermal Synthesis of Silver Nanoprisms with Chemically Tailorable Thickness. *Adv. Mater.* 2005; 17: 412–415.

14. Wiley B, Sun Y, Xia Y. Synthesis of Silver Nanostructures with Controlled Shapes and Properties. *Acc. Chem. Res.* 2007; 40: 1067–1076.

15. Cathcart N, Frank AJ, Kitaev V. Silver Nanoparticles with Planar Twinned Defects: Effect of Halides for Precise Tuning of Plasmon Resonance Maxima from 400 to > 900 *Nm. Chem. Commun.* 2009; 46: 7170–7172.

## *Procedure*

### I.  Diffraction from Hair

For this experiment, we will generate a diffraction pattern by placing a human hair in the beam of a laser pointer. This experiment is best done in pairs, with one person holding and manipulating the laser pointer and the other person making the distance measurements.

**WARNING: Do not shine the laser pointer into anybody's eyes!**

1. Remove a single strand of hair from your head; tape the hair across the end of the laser pointer, being sure to place the hair directly in the path of the beam.

2. Shine the laser pointer directly at the wall or a piece of paper. Try to have the beam be perpendicular to the wall/paper surface. You should see a bright point of light, with a fainter diffraction pattern spreading to each side (Figure 6; you can rotate the pointer to line the diffraction pattern up either horizontally or vertically on the surface). The farther you are from the wall, the larger the pattern will be, and the easier it will be to see and measure the diffraction pattern—a distance of 4 to 6 ft generally works well.

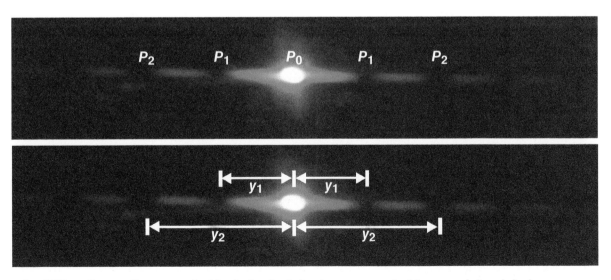

**Figure 6**  **Diffraction pattern created by hair stretched across end of laser pointer; labels are described in the text.**

3. Measure the distance $L$ from the front of the laser pointer to the center of the bright point on the surface ($P_0$). To obtain a fixed accurate distance, it is important to hold the laser beam in a fixed position. If necessary, the laser pointer can be taped down to the surface of a lab table, pointing toward a wall, to keep the value of $L$ fixed.

4. Measure the distances $y_1$ and $y_2$ from the center of the bright point to the *centers* of the first two dark bands on one side of the diffraction pattern ($P_1$ and $P_2$).

5. Repeat steps 1 through 4 using a hair sample from the other lab partner.

## II.  Synthesis of Silver Nanoprisms (be sure all glassware is clean)

1. Label three large test tubes with the numbers "1.0", "1.5", and "2.0", respectively.
2. To each of the tubes add the following solutions:
   - First, 5.0 mL of the stock sodium citrate solution
   - Second, 5.0 mL of the stock $AgNO_3$ solution
   - Third, 3.0 mL of the stock $H_2O_2$ solution
3. Shake the tubes to mix the solutions.
4. Then, add different volumes of the stock KBr solution and $H_2O$ to each of the tubes. *The KBr volume added is critical to the size of the nanoprisms synthesized, so precision is important.* The following volumes are added:
   - 1.0 mL KBr and 1.0 mL $H_2O$ to the tube labeled "1.0"
   - 1.5 mL KBr and 0.5 mL $H_2O$ to the tube labeled "1.5"
   - 2.0 mL KBr to the tube labeled "2.0" (no $H_2O$)
5. Again, shake the tubes to mix.
6. Finally, add 2.5 mL of the stock $NaBH_4$ solution to the *first tube only* [the borohydride is used to reduce the $Ag^+$ to $Ag(s)$]. Then seal the tube with a stopper or parafilm, place your thumb over the end to seal, and invert the tube three times to mix (be sure covering is secure to avoid skin contact)—**do not shake vigorously!** Repeat this step for the other two tubes (again, one at a time).
7. A color change should occur in each vial and be complete after about 3 min. Record the color of each solution.
8. Obtain a small sample (~5 mL) of an "Unknown" nanoprism sample from your instructor.
9. Identify the approximate wavelength of maximum absorbance ($\lambda_{max}$) for each sample based on their color by looking at Figures 3 and 4; for example, for a yellow solution, the solution is yellow because the complement to yellow, or violet, is being absorbed; so the absorbance max is approximately 400 nm.
10. Using your spectrophotometer, determine the absorbance maximum for each nanoprism solution that you synthesized. Repeat for each nanoprism sample; if you are using the same cuvette for each sample, rinse before adding each new sample.
11. Record the color and also determine the absorbance maximum for the "Unknown" nanoprism solution provided to you by your instructor.

## Data Collection

### I.  Diffraction from hair

Hair sample 1:

$L$ _____

$y_1$ _____

$y_2$ _____

hair color _____

Hair sample 2:

$L$ _____

$y_1$ _____

$y_2$ _____

hair color _____

## II. Synthesis of Silver Nanoprisms

| Nanoprism Sample | Color | Estimated $\lambda_{max}$ (nm) | Measured $\lambda_{max}$ (nm) |
|---|---|---|---|
| "1.0" | | | |
| "1.5" | | | |
| "2.0" | | | |
| | | | |
| Unknown sample | | | |

## Analysis

### DIFFRACTION FROM HAIR

1. Calculate the values of $\theta_1$ and $\theta_2$ for each hair sample using Equation 5; show your calculations:

**Hair sample 1:**

$\theta_1 = $ _____

$\theta_2 = $ _____

**Hair sample 2:**

$\theta_1 = $ _____

$\theta_2 = $ _____

**2.** Calculate the width of each hair ($w$) using Equation 4. For each sample you will generate two values (one based on $\theta_1$ and another based on $\theta_2$). Then calculate the average width for each hair; show your calculations:

**Hair sample 1:**

$w$ (based on $\theta_1$) = _____

$w$ (based on $\theta_2$) = _____

$w_{average}$ = _____

**Hair sample 2:**

$w$ (based on $\theta_1$) = _____

$w$ (based on $\theta_2$) = _____

$w_{average}$ = _____

# SILVER NANOPRISMS

1. On the graph below, plot $\lambda_{max}$ for each synthesized nanoprism sample ($y$-axis) versus the expected nanoparticle size ($x$-axis; from "Average Width" column in Table 2). Then, use a ruler to draw a "best-fit" line to the data (the best-fit line should pass as close as possible to all three data points):

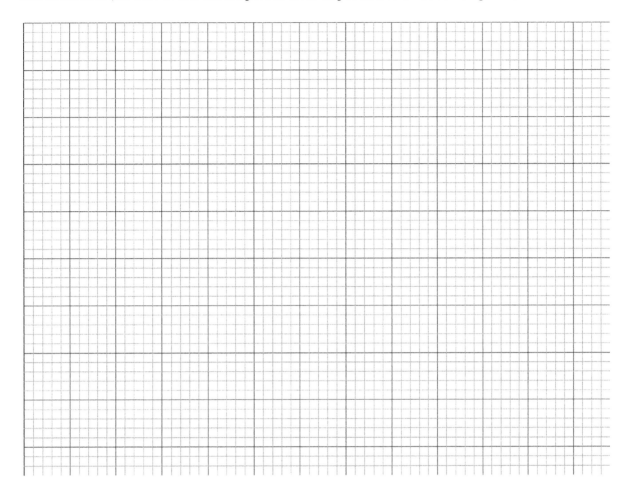

2. Using the best-fit line you have just drawn, estimate the nanoprism size associated with the unknown nanoprism sample (use the $\lambda_{max}$ you measured for that sample):

   nanoprism size: _____

## Reflection Questions

1. If you were to switch to a green laser pointer ($\lambda = 532$ nm) in the diffraction experiment, describe what you would expect to change when you conduct the experiment (assume you stand the same distance from the wall):

2. List the possible sources of error in the diffraction experiment. How would you address each one?

3. What is the color of elemental silver—explain your answer:

4. Which color of light is diffracted at a greater angle from a diffraction grating, red or yellow light?

5. For a substance to look red, describe how it interacts with white light.

6. If the color of a silver nanoparticle solution remains constant, what does that tell you about the tendency of the particles to aggregate under those conditions?

## Connection

Based on your experience in this lab, draw a connection to something in your everyday life or the world around you (something not mentioned in the background section).

# 4

# Shedding Light on the Structure of the Atom

## Objectives

- Build and calibrate a simple spectroscope.
- Use your spectroscope to observe the line spectrum of sodium; calculate the energy and frequency associated with the major transition(s).
- Use your spectroscope to observe the line spectra of unknown substances; use line spectrum to identify substance.

## Materials Needed

- A one-piece box (8" × 8" × 2") [versions can be purchased at The Container Store (item 162004), or online from various sources including FetPak Inc. (http://www.fetpak.com; item AW22) and Paper Mart (http://www.papermart.com; item 4011480)]
- Holographic diffraction grating (slide, or 24 mm × 36 mm piece of paper); 1000 lines/mm preferred [multiple sources online include Rainbow Symphony Store (http://www.rainbowsymphonystore.com; item 01504 or 01603)]
- Aluminum foil to make slit
- Black tape (duct, electrical)
- Scotch tape
- Ruler
- Scissors
- Something to cut box (e.g., razor blade, box cutter, or exacto knife)
- Light sources for observation (e.g., gas discharge tubes filled with various gases)

## Safety Precautions

- **Never look directly at sunlight or a laser light (e.g., a laser pointer) with your spectroscope!** If you are using one of these as your light source, look at it indirectly [as it reflects off of a white surface (e.g., wall, sheet of paper, etc.)].
- Some light sources (e.g., hydrogen and nitrogen lamps) may emit ultraviolet radiation, which can damage your eyes. The typical safety glasses will protect your eyes from the UV radiation, so **keep your safety glasses on at all times.**
- If you are working with gas discharge tubes, be sure to unplug the power supply before removing or adjusting the position of the tube.
- If you are working with open flames as source of light, be sure not to catch your spectroscope on fire.

## Background

Because atoms are too small to be observed directly, scientists in the late nineteenth and early twentieth centuries had to use indirect methods to decipher the architecture of individual atoms. These classic and creative experiments, which included cathode ray tubes, Millikan's oil drop experiment, and Rutherford's gold foil experiment, led to the atomic model with a nucleus containing protons and neutrons surrounded by a relatively large "cloud" of electrons.

But are the electrons further organized around the nucleus, or is the implied randomness in the preceding "cloud" analogy accurate? Again, scientists have been forced to utilize indirect methods to understand how electrons are arranged in an atom. For this question, **light** has been utilized to probe the electronic structure of atoms. Specifically, scientists have used light (or more accurately electromagnetic radiation), and the absorption and emission of light by matter, to determine that indeed electrons *do* have a specific organization in each and every atom.

The classic (and most simple) example for us to consider is hydrogen, which has one proton in the nucleus and one electron. When hydrogen gas is sealed inside a glass tube and submitted to an electric current, we observe a reddish glow emanating from the tube. [Note: Light is emitted when any matter absorbs enough energy: hot objects generate light (e.g., piece of metal in a fire), gases emit light when an electric current is passed through them, and some forms of matter will burn with a noticeable color when placed in a flame.] When we pass this light through a slit and a prism or a diffraction grating (see the previous lab), we observe a series of individual lines of different colors and *not* a continuous spectrum with all wavelengths present. Interestingly, we *always* see the exact same lines for hydrogen—never more or less, and never at different wavelengths (colors in the visible region of the electromagnetic spectrum). This series of lines is called an atomic spectrum or line spectrum, and it is diagnostic for the presence of hydrogen; in other words, different elements will generate different line spectra (each unique for that specific element). The lines for hydrogen that are observed in the visible region have been well characterized (see Table 1 and Figure 1).

| Table 1 Line Spectrum Values for Hydrogen in the Visible Region.[1] | |
| --- | --- |
| **Wavelength (nm)** | **Color** |
| 383.5384 | Violet |
| 388.9049 | Violet |
| 397.0072 | Violet |
| 410.174 | Violet |
| 434.047 | Violet |
| 486.133 | Blue-green |
| 656.272 | Red |
| 656.2852 | Red |

**Figure 1  Line spectrum for hydrogen.**

This observed spectrum generates several questions:

- How are these lines generated?
- How can there be multiple lines if hydrogen only has one electron?
- Why are these lines at these specific wavelength values? Why not different wavelength values? Why not more or fewer lines?
- Why are line spectra different for different elements?

Most importantly, any successful model of the atom must provide an understanding and explanation of the line spectral data.

Based on the work of Niels Bohr,[2–4] we have a simple model for the atomic structure of the atom that can help explain the line spectrum of hydrogen. In this model, electrons ($e^-$) in an atom are located in fixed orbits that encircle the nucleus. An $e^-$ in a given orbit has a specific energy given by Equation 1:

$$E = -\frac{2\pi^2 m e^4}{n^2 h^2} = -\frac{b}{n^2} \qquad \textbf{Equation 1}$$

where $m$ is the $e^-$ mass, $e$ is the $e^-$ charge ($-1.6022 \times 10^{-19}$ C), $h$ is Planck's constant ($6.63 \times 10^{-34}$ J s), and $b$ = $2.178 \times 10^{-18}$ J (combination of all constants in the equation). The variable $n$ can have values 1, 2, . . . and is known as the **principal quantum number.** Each orbit has a different value of $n$; the lowest orbit (closest to the nucleus) has $n = 1$, the next farthest from the nucleus has $n = 2$, and so forth. An $e^-$ located in the first Bohr orbit ($n = 1$) has the lowest (most negative) energy because it is closest to the nucleus. As the value of $n$ increases, an $e^-$ will be located farther from the nucleus and therefore have a higher (more positive) energy. Another important tenet of Bohr's model is that the $e^-$ can only have an energy associated with one of the orbits; that is, the $e^-$ can never have an intermediate energy. Thus, we say that the energy of an $e^-$ is **quantized.**

Using this model, hydrogen has one $e^-$ located in the $n = 1$ Bohr orbit. This $e^-$ is able to move away from the nucleus by **absorbing** a specific amount of energy and relocating to a new orbit with a larger $n$ value (position of higher potential energy)—this is known as an **excited state** of the atom, and it is less stable than the original **ground state** (lowest energy possible). The difference between the energies of the two orbits between which the $e^-$ moves reflects the amount of energy absorbed by the $e^-$. This difference can be calculated using Equation 2:

$$\Delta E = E_f - E_i = -b\left(\frac{1}{n_f^2} - \frac{1}{n_i^2}\right) \qquad \textbf{Equation 2}$$

Here the subscripts $i$ and $f$ stand for initial and final states of the $e^-$, respectively. Being unstable, the $e^-$ will eventually release a specific amount of energy in the form of light (**emission**) and return to a lower energy orbit, with $\Delta E$ calculated using Equation 2. Bohr reasoned that the lines observed in the atomic spectrum of hydrogen arise from the $e^-$ transitions between the different orbits. Since the orbit energies are quantized, $\Delta E$ can have only specific values, and therefore only specific lines are observed with specific wavelengths (colors) associated with them.

$$c = \lambda\nu \qquad \Delta E = h\nu \qquad \Delta E = (hc)/\lambda \qquad \textbf{Equations 3a–c}$$

where

$c$ = speed of light = $2.998 \times 10^8$ m s$^{-1}$
$h$ = Planck's constant = $6.63 \times 10^{-34}$ J s (photon)$^{-1}$
$\nu$ = frequency (s$^{-1}$ or Hz)
$\lambda$ = wavelength (m)

Although the simple Bohr model works only for hydrogen and any other one $e^-$ ion (the model is incomplete for multielectron systems because it does not account for interactions between the electrons), we can still use Equations 3a–c to relate the wavelength of a given line in a line spectrum of any element to the $\Delta E$ that an $e^-$ undergoes.

The line spectrum of hydrogen has been heavily studied for a number of reasons. With only one electron involved, it should be the easiest to interpret. Also, it is useful in astronomy, since these lines show up in the spectra associated with stellar objects in the universe that are commonly studied based on their emitted light. The Balmer series for hydrogen[5] is the series of lines that result from the $e^-$ relaxing from various final orbits ($n_f$) to the $n = 2$ orbit. We can modify Table 1 to include the orbit-to-orbit transitions that are specifically involved for each line, generating Table 2.

**Table 2    Balmer Series for Hydrogen in the Visible Region.[5]**

| Wavelength (nm) | Color | Transition ($n_f \rightarrow n_i$) |
| --- | --- | --- |
| 383.5384 | Violet | $9 \rightarrow 2$ |
| 388.9049 | Violet | $8 \rightarrow 2$ |
| 397.0072 | Violet | $7 \rightarrow 2$ |
| 410.174 | Violet | $6 \rightarrow 2$ |
| 434.047 | Violet | $5 \rightarrow 2$ |
| 486.133 | Blue-green | $4 \rightarrow 2$ |
| 656.272 | Red | $3 \rightarrow 2$ |
| 656.2852 | Red | $3 \rightarrow 2$ |

In this lab, the goal is to create a spectroscope that will use a diffraction grating to separate light emission into a rainbow (for a continuous light source) or a line spectrum for elemental sources; numerous variations on the spectroscope design used here can be found in References 6–12. You will then calibrate the spectroscope using the fluorescent lights in the lab as your light source. Finally, you will use the spectroscope to measure the transition energies for various elements as well as to measure line spectra for unknown samples to determine their identities.

## References

1. *CRC Handbook of Chemistry and Physics*. 53rd ed. (Weast RC, Ed.). 1972; The Chemical Rubber Co, Ohio.
2. Bohr N. On the Constitution of Atoms and Molecules, Part I. *Philosophical Magazine*. 1913; 26: 1–24.
3. Bohr N. On the Constitution of Atoms and Molecules, Part II. Systems Containing Only a Single Nucleus. *Philosophical Magazine*. 1913; 26: 476–502.
4. Bohr N. On the Constitution of Atoms and Molecules, Part III. Systems Containing Several Nuclei. *Philosophical Magazine*. 1913; 26: 857–875.
5. Birge RT. The Balmer Series of Hydrogen, and the Quantum Theory of Line Spectra. *Phys. Rev.* 1921; 17: 589–607.
6. Edwards RK, Brandt WW, Companion AL. A Simple and Inexpensive Student Spectroscope. *J. Chem. Ed.* 1962; 39: 147–148.

7. Wakabayashi F. Resolving Spectral Lines with a Periscope-Type DVD Spectroscope. *J. Chem. Ed.* 2008; 85: 849–853.
8. Wakabayashi F, Hamada K, Sone K. CD-Rom Spectroscope: A Simple and Inexpensive Tool for Classroom Demonstrations on Chemical Spectroscopy. *J. Chem. Ed.* 1998; 75: 1569–1570.
9. Cortel A, Fernandez L. A Simple Diffraction Grating Spectroscope: Its Construction and Uses. *J. Chem. Ed.* 1986; 63: 348–349.
10. Westra MT. A Fresh Look at Light: Build Your Own Spectrometer. *Science in School.* 2007: 30–34.
11. Briggs RP, Carlisle RJ. *Building a Spectroscope.* In *Solar Physics and Terrestrial Effects: A Curriculum Guide for Teachers Grades 7–12.* 2nd ed. (Poppe BB, Ed.), pp. 104. 1996; Space Environment Center, National Oceanic and Atmospheric Administration.
12. Website www.swpc.noaa.gov/Curric_7-12/Activity_1.pdf (last accessed July, 2013).

## Procedure

### I. Construct Spectroscope

1. Fold an 8 in × 8 in × 3.5 in (203 mm × 203 mm × 89 mm) box.
2. Use a strip of black tape on the inside of the box to secure the bottom fold and prevent light from leaking through.
3. On the bottom face of the box, use your ruler to draw a line that splits the box bottom in half; this is the long leg of a right triangle (connects points A and B in Figure 2).
4. To generate the correct angle (15°) between the slit and the diffraction grating, the short leg of the triangle (line B–C in Figure 2) will measure 53 mm long. Note that the length of leg B–C will be different if you use a box that has a length (leg A–B) different than 8 in; to calculate, use Equation 4.

$$\tan(\theta) = \frac{B - C}{A - B}$$

**Equation 4**

Here $\theta$ = angle C–A–B = 15°, and A to B and B to C are the lengths of the two legs of the triangle (Figure 2).

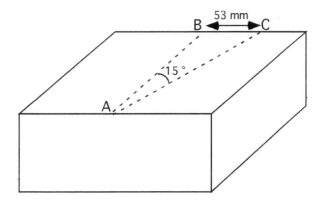

**Figure 2**

5. Incorporate the slit (Figure 3):
   - Cut a 20 mm (wide) × 40 mm (high) rectangular hole centered above point C.
   - Place the piece of aluminum foil centered over the hole and tape into place.
   - Use a razor blade or something sharp to cut a very thin (< 0.5 mm) vertical slit in the foil.

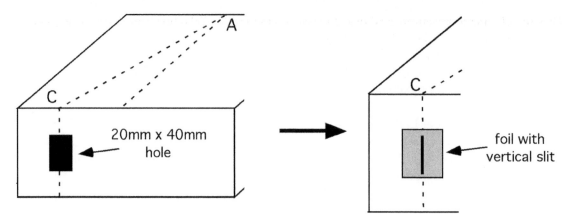

**Figure 3**

6. Incorporate the diffraction grating (Figure 4):
   - On the end of the box above point A, cut a 34 mm (wide) × 24 mm (high) rectangular hole, centered above A.
   - Take a piece of diffraction grating paper 50 mm (wide) × 40 mm (high), put a piece of Scotch tape on one edge.
   - On the inside of the box, place the grating over the hole, square with the sides of the hole.
   - Close the box, point the slit directly at the fluorescent lights in the room, and look for a line spectrum to the right of the slit inside the box.
   - If a spectrum does not appear, then your grating needs to be rotated 90° to align the grating lines with the slit orientation.
   - Once properly aligned, use Scotch tape to tape down the other three edges of the grating.

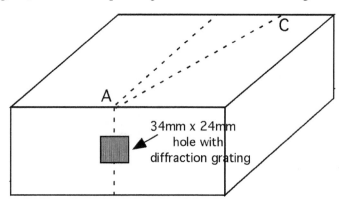

**Figure 4**

7. Incorporate the grid for spectral measurements:
   - Use a pencil to mark inside the box approximately where the ends of the spectrum appear.
   - Use a razor blade to cut a flap that overlaps this location but is displaced toward the bottom of the box (see the illustration for the approximate configuration); the flap should be hinged at the *top* (toward the middle of the box), and should open from the *bottom*; the goal is to have the hole that you've cut slightly overlap the line spectrum; the hole should be ~60 mm (wide) × ~15 mm (high).
   - Copy this grid from lab manual:

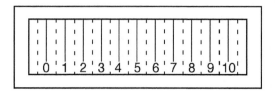

- Use Scotch tape to attach the grid over the hole on the inside of the box; the numbers and lines should point to the inside of the box (Figure 5).

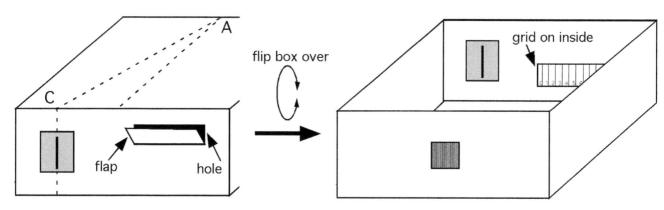

**Figure 5**

8. You can now close the box and, if too much light leaks in, seal the box shut using black tape. The goal is to have the slit, the diffraction grating, and the measurement grid window be the only sources of light inside the box.

## II. Calibrate Spectroscope

To make the spectroscope useful as a tool for quantitative measurements, we must calibrate it using a light source that generates a well-characterized line spectrum with precisely known wavelengths. The fluorescent lights in your lab contain mercury vapor and work well for calibration purposes. The known wavelength values for observed lines in the Hg line spectrum are listed in Table 3.[1]

| Table 3  Line Spectrum Values for Mercury in the Visible Region.[1] | |
|---|---|
| **Color** | **Wavelength (nm)** |
| Violet | 404.7 |
| Blue | 435.8* |
| Green | 546.1* |
| Yellow-orange | 577.0* |
| Yellow-orange | 579.1* |
| Orange-red | 623.4 |
| Red | 690.7 |

\* Most intense lines

1. Point the slit of your spectroscope toward the fluorescent lights in the lab while looking through the diffraction grating; a line spectrum should appear to the right of the slit.
2. Adjust the grid opening so that you can see both the line spectrum and the grid; for each line, record the color and its position on the numbered grid on the data sheet.
3. Use the grid position versus known wavelength data to generate a best-fit (regression) line for use as a calibration curve (see the Analysis section).

## III. Measurements on Sodium Sample and Unknown Sample(s)

1. Using the procedure just described, record the color and position of each line for the sodium sample. Sodium has a prominent line called the sodium D line. This line is due to the valence electron in Na moving from an excited state (a $3p$ orbital) back to its ground state $3s$ orbital.
2. Record the color and position of each line for each unknown sample.

## Data Collection

1. Calibration data for spectroscope:
   - Record grid positions in this table.

| Color | Wavelength (nm) | Grid Position |
|-------|-----------------|---------------|
| Violet | 404.7 | |
| Blue | 435.8* | |
| Green | 546.1* | |
| Yellow-orange | 577.0* | |
| Yellow-orange | 579.1* | |
| Orange-red | 623.4 | |
| Red | 690.7 | |

\* Most intense lines

2. Line spectrum data for sodium sample:
   - Draw a line on the grid below in the correct position for each observed line.
   - Above each line, indicate the color of the line.
   - Below each line, indicate the grid position numerical value.

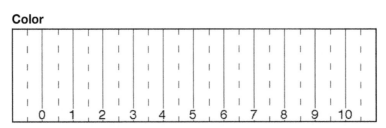

3. Line spectrum data for unknown sample(s): for each sample,
   - Draw a line on the grid below in the correct position for each observed line.
   - Above each line, indicate the color of the line.

- Below each line, indicate the grid position numerical value (e.g., if the line is between 2 and 3 on the grid, you might write "2.4" for the grid position).

**SAMPLE 1**

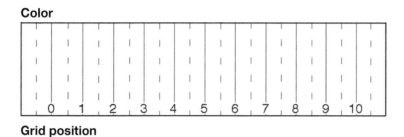

**SAMPLE 2**

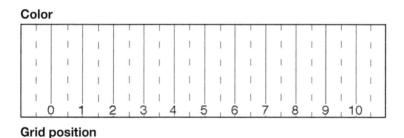

**SAMPLE 3**

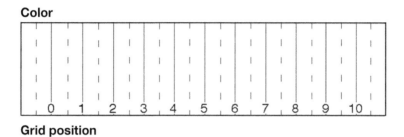

**SAMPLE 4**

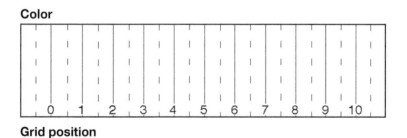

## Analysis

1. Generation of calibration curve:
   - To generate the line that best fits our calibration data, we will be using the method of least squares. The basic idea is to minimize the distance of the line from all the data points concurrently. The calculus behind this method (which involves partial derivatives) is beyond the scope of this lab, yet the calculations are straightforward enough.

- The line $y = mx + b$, the least-squares or best-fit line that is associated with the data points $(x_1, y_1)$, $(x_2, y_2)$, . . . , $(x_n, y_n)$, can be calculated using the following equations:

slope: 
$$m = \frac{n(\sum xy) - (\sum x)(\sum y)}{n(\sum x^2) - (\sum x)^2}$$
**Equation 5**

y-intercept: 
$$b = \frac{\sum y - m(\sum x)}{n}$$
**Equation 6**

where

$$\sum xy = x_1 y_1 + x_2 y_2 + \cdot\cdot\cdot + x_n y_n$$

$$\sum x = x_1 + x_2 + \cdot\cdot\cdot + x_n$$

$$\sum y = y_1 + y_2 + \cdot\cdot\cdot + y_n$$

$$\sum x^2 = x_1^2 + x_2^2 + \cdot\cdot\cdot + x_n^2$$

**(a)** Fill in the following table based on the calibration data you collected:

| x (wavelength) | y (grid position) | xy | x² |
|---|---|---|---|
| 404.7 | | | |
| 435.8* | | | |
| 546.1* | | | |
| 577.0* | | | |
| 579.1* | | | |
| 623.4 | | | |
| 690.7 | | | |

* Most intense lines

**(b)** Use the values from part (a) to calculate the following values:

$$\sum x \; = \underline{\hspace{3cm}}$$

$$\sum y \; = \underline{\hspace{3cm}}$$

$$\sum xy = \underline{\hspace{3cm}}$$

$$\sum x^2 = \underline{\hspace{3cm}}$$

**(c)** Plug the values from part (b) into Equations 5 and 6 to calculate the slope and $y$-intercept of the least squares line [$n$ is the number of $(x, y)$ data points from part (a)].

$$m = \underline{\hspace{3cm}}$$

$$b \; = \underline{\hspace{3cm}}$$

**(d)** Using your data in part (c), plot your line spectrum data and least-squares line on the following graph:

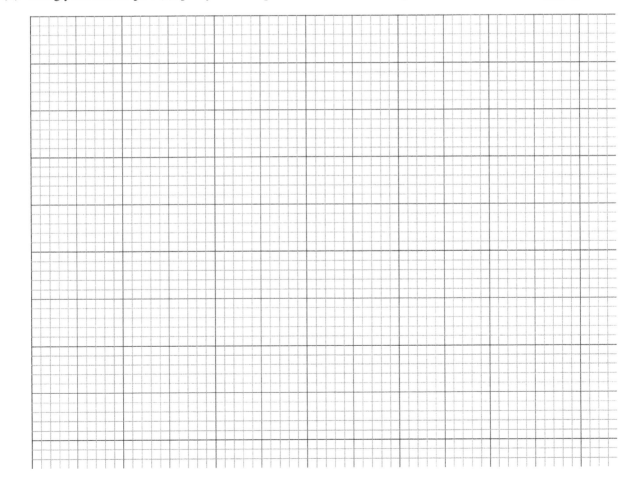

**2.** Calculate the energy associated with the line from the sodium sample:

(a) Use your calibration curve to determine the wavelength of the sodium line you observed. Show your calculations:

$\lambda =$ _____

(b) The literature value for the wavelength associated with the sodium line is 589 nm. Calculate the percent error associated with the wavelength you calculated, where the percent error is calculated using Equation 7:

$$\% \text{ error} = \frac{(\text{measured}) - (\text{literature})}{\text{literature}} \times 100 \qquad \textbf{Equation 7}$$

$\% \text{ error} =$ _____

(c) Convert the wavelength you generated in part (a) to the energy associated with this transition. Show your calculations:

$\text{energy} =$ _____

**3.** Identification of unknowns using line spectra:
- For each unknown, compare the line spectrum you recorded above to the set of known line spectra in Figure 6 to identify each substance:

| Substance | Identity (based on line spectra) |
|-----------|----------------------------------|
| SAMPLE 1  |                                  |
| SAMPLE 2  |                                  |
| SAMPLE 3  |                                  |
| SAMPLE 4  |                                  |

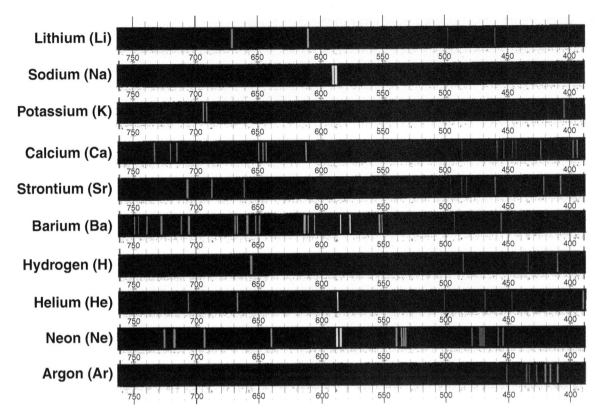

Figure 6 Line spectra of various elements.

## Reflection Questions

1. The spectral lines observed for hydrogen arise from transitions from excited states back to the $n = 2$ principal quantum level. Calculate the wavelengths associated with the spectral transitions of the hydrogen atom from the $n = 6, 5, 4,$ and $3$ levels to the $n = 2$ level.

2. Suppose you use a diffraction grating that contains 600 lines mm$^{-1}$ in your spectroscope. What effect would you expect this to have on the sharpness and resolution of the lines in your line spectrum?

3. Let's imagine that you were going to start a company that produces spectroscopes to sell to general chemistry students. How could you ensure that all the spectroscopes created the same calibration line and equation? What would you need to be sure was always the same? What could be changed without affecting the calibration?

4. What are the possible sources of error in the wavelength of the sodium line you calculated in Analysis part 2(b)?

5. As indicated earlier, white light contains a continuous spectrum including all the visible wavelengths of light. Interestingly, if one uses a spectroscope to look at sunlight (**not directly at the sun, which would cause damage to your eyesight!!**, but indirectly by pointing the spectroscope at a clear part of the sky), we observe numerous dark lines superimposed on the continuous spectrum. This phenomenon was originally observed by Wollaton in 1802 and then again later by Fraunhofer in 1817. Called Fraunhofer lines, these lines correlate with the line spectra of various elements that are thought to exist in the outer atmosphere of the sun (Figure 7).

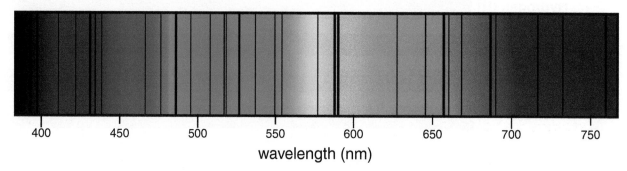

**Figure 7  Fraunhofer lines.**

(a) Explain the process by which these dark lines are generated by the elements in the sun's atmosphere. Specifically, why are they black and not colored, and why do they match the colored line spectra of those elements?

(b) Compare the Fraunhofer lines to the line spectra in Figure 6. Which of these elements are in the sun?

## Connection

Based on your experience in this lab, draw a connection to something in your everyday life or the world around you (something not mentioned in the background section).

# 5

# Periodic Trends: Densities in the Chromium Family of Transition Metals

## Objectives

- Measure the bulk densities and calculate atomic densities of three transition elements in the same family of the periodic table.
- Relate measured density to atomic size, a periodic trend.

## Materials Needed

- Chromium (Cr), 25 to 30 g
- Molybdenum (Mo), 35 to 40 g
- Tungsten (W), 55 to 60 g
- Forceps or tongs
- Water
- 25-mL (or 10-mL) graduated cylinder

## Safety Precautions

- Gloves and safety goggles/glasses should be worn during this lab.

## Background

Patterns in the properties of the elements have been apparent to scientists since the early 1800s, serving initially as a means to their organization. Indeed, the German chemist Johann Döbereiner identified "triads" of elements with similar chemical properties as early as 1829 (the triads, such as Li—Na—K and S—Se—Te, are family members in the modern periodic table).[1] In the mid-1860s, the "law of octaves" was coined by English chemist and music-lover John Newlands, the name referring to the repeating pattern in chemical properties for every eighth element (when arranged in order of increasing atomic mass).[2] Around the same time (1864 to 1869), Dmitri Mendeleev (Russia) and Lothar Meyer (Germany) independently utilized ideas of periodicity to create charts with the elements grouped in vertical columns, reflecting similar properties, and again arranged by increasing atomic mass. Finally in 1913, the English physicist H. G. J. Moseley modified these charts, using atomic number instead of mass to order the elements, giving rise to the modern periodic table.[3] As we know now, it is the number and arrangement of *electrons*, equal to the atomic number in a neutral atom, that gives rise to atomic properties (and chemical reactivity), thereby dictating the location of each element in the periodic table.

Inspection of the periodic table reveals clear trends in atomic properties, where the property depends on the location of the element either in a family (vertical column) or within a period (horizontal row). For example, as one progresses down the alkali metal family (Group IA), each atom becomes larger and more readily loses a single electron to form a "+1" cation, the common charged atom for group IA members. Likewise, a comparison of the second-period elements Li and F shows that fluorine is smaller and loses electrons less readily than lithium. In fact, atomic fluorine readily gains a single electron to form a "−1" anion, a property common to all members of the halogen family (Group VIIA). A simplified view of the attraction between an atom's electrons and the nucleus provides a means for understanding these periodic trends. Thus, progressing sequentially across a period (left to right), a single proton and electron are added to each successive atom, with the electron placed into the same shell (valence shell) for the main group elements. Electrons in the inner orbitals shield or screen the outermost electrons from feeling the full nuclear charge ($Z$, which is equivalent to atomic number). This leads to an effective nuclear charge ($Z_{eff}$), or reduced magnitude of positive charge "experienced" by the outermost electrons in the atom. In general, the value of $Z_{eff}$ increases as one proceeds left to right across a period, reflecting incomplete shielding which, in turn, creates an increase in the attraction of the outermost electrons toward the nucleus. In contrast, $Z_{eff}$ decreases moving down a group as outermost electrons are placed in orbitals that are successively farther from the nucleus. The change in $Z_{eff}$ results in predictable trends in important atomic properties:

*Trend 1. Atomic radius:* (Figure 1) The distance from the center of the nucleus to the outermost electrons decreases as you move from left to right in a period due to increasing $Z_{eff}$.

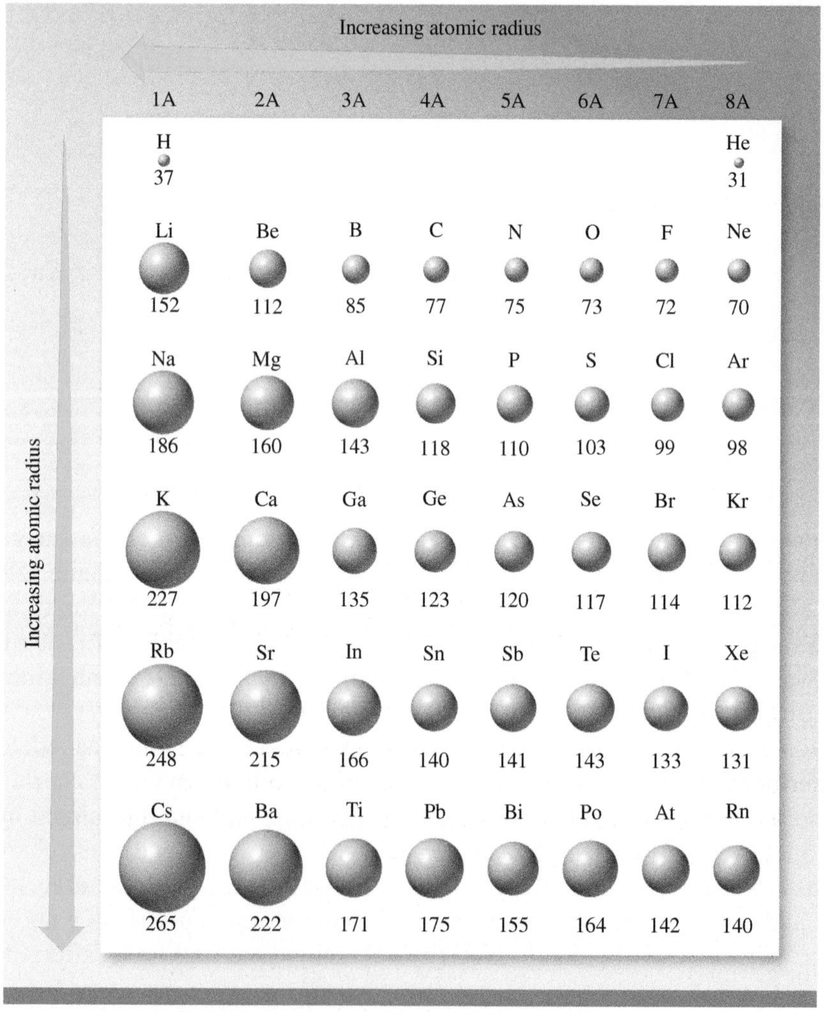

**Figure 1    Atomic radii of representative elements (listed in picometers).**

*Trend 2. Ionization energy (IE):* (Figure 2) The minimum energy required to remove one electron from a gaseous atom/ion in its ground state electron configuration (kJ mol$^{-1}$).

$$X(g) \rightarrow X^+(g) + e^-$$

*IE* generally increases as you proceed left to right in a period; the increasing $Z_{eff}$ leads to an increased hold on the outermost electrons. IE correlates with the reactivity of a metal—the smaller the *IE*, the more reactive the metal (easier to remove an electron).

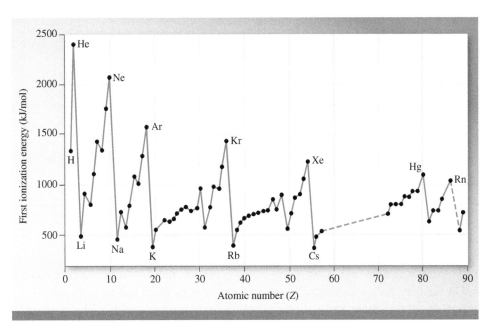

**Figure 2** **First ionization energies vs. atomic number.**

*Trend 3. Electron affinity (EA):* (Figure 3) The energy change associated with the addition of an electron to a gaseous atom/ion (kJ mol$^{-1}$).

$$X(g) + e^- \rightarrow X^-(g)$$

*EA* generally increases as you proceed left to right in a period. *EA* correlates with the reactivity of a nonmetal—the larger the *EA*, the more reactive the nonmetal (the easier to add an electron).

Exceptions are observed in the *IE* and *EA* general trends due to the extra stability of half-filled and completely filled subshells (e.g., the first *IE* of nitrogen is larger than that of oxygen). Exceptions are best predicted or explained by examination of electron configurations using orbital diagrams.

As one moves down a family, electrons are placed in higher energy orbitals (greater "n" value), which are farther from the nucleus with greater shielding provided by the increased number of closed shell core electrons. The following trends are observed as a result:

*Trend 1. Atomic radius:* The atomic radius increases progressing down the family.

*Trend 2. Ionization energy:* IE decreases progressing down a family since the outermost electron that is being removed is farther from the nucleus and thus feels a weaker attraction.

*Trend 3. Electron affinity:* EA also decreases progressing down a family because the added electron is placed in an orbital that is, on average, farther from the nucleus.

Interestingly, density (mass per volume with units of g cm$^{-3}$ or g mL$^{-1}$) also displays a notable periodic trend. As a matter of fact, density was one of the physical properties utilized by both Mendeleev and Meyer (along with atomic mass, melting point, and specific heat) when deriving their original periodic charts. As one goes down a family, the atomic density (density of an atom) increases—its mass increases (based primarily on the mass of the nucleus, since electrons do not contribute significant

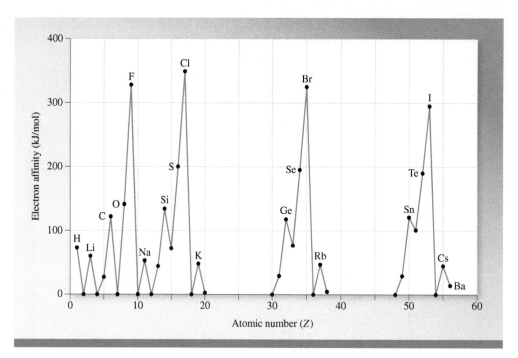

**Figure 3  Electron affinities vs. atomic number.**

mass). The volume of an atom also increases going down a family, but not to the same extent as the increase in mass. The density varies less significantly as one moves from left to right across a period because the variations in atomic mass and volume between neighboring atoms within a period are rather small. As shown in Figure 4, a comparison of two similarly sized objects (golf balls) with differing masses provides a visual demonstration of density (because atoms are spherical in shape, golf balls serve as convenient macroscopic analogs). The two golf balls shown occupy the same volume, but they differ significantly in mass. The plastic golf ball is hollow and the authentic golf ball contains a solid core. When both are placed in a container of water, the plastic ball floats and the authentic one sinks, demonstrating that the density of the latter is larger than that of the plastic ball (and, also of water, as any golfer knows!).

**Figure 4  Plastic and real golf balls in a beaker of water.**

The density that is typically reported for a substance is not per atom, but for a large number of atoms—that is, density is a bulk property. As such, one must consider how atoms pack together to form the bulk sample to understand the density of a substance. For pure metals, atoms are arranged in lattices that contain one of four repeating patterns (unit cells). Figure 5 shows the body-centered cubic (BCC) packing arrangement utilized by the elements you'll be using in this lab. The BCC packing is formed with an alternating AB pattern (ABABABABAB . . .), where layer A forms a square-packed plane in which the atoms do not quite touch each other, and layer B packs above A with atoms sitting in the holes or depressions created by layer A. Continuing with the golf ball analogy, if both types of golf balls were put separately into BCC packing arrangements, they would each occupy the same bulk volume but have different masses (and therefore different bulk densities). Different densities could also be generated by changing the packing arrangement. *Importantly, in the context of real solid samples, as long as the solids that are being compared have the same packing arrangement, the atoms will be arranged identically, and the bulk densities of the solids can be directly compared to learn about atomic properties.*

Body-centered cubic

**Figure 5   Illustration showing the body-centered cubic (BCC) packing arrangement.**

## THIS LAB

In this lab, you will first use literature values to calculate the atomic densities of three transition metals. Then, you will use accurate measurements of volume and mass for the three metals to calculate their bulk densities, ultimately comparing atomic density to bulk density to better understand their periodic properties. The metals selected for this lab impact our everyday lives, but are probably not as familiar to you as others.

*Chromium:* a silver and lustrous metal that is hard and exhibits high corrosion resistance. Chromium is used for chrome plating. Chromium in the $Cr^{3+}$ oxidation state is required by biological systems for sugar and lipid metabolism. Finally, chromium is part of the natural yellow pigment chrome yellow ($PbCrO_4$); this was used for years in the paint for school buses, but has recently been replaced with organic dyes due to health concerns regarding the lead and the hexavalent chromium species.

*Molybdenum:* a silvery-white metal whose name comes from the Greek word "molybdos," which means lead (it is similar in appearance to and was originally mistaken for lead). Molybdenum has a very high melting point (2623°C) and is used in high-strength alloy molybdenum steel. The metal is also used in the petroleum industry for removal of sulfur compounds in coal and gas liquification processes. Molybdenum is the only second row transition metal that is required by a number of living systems. It is an essential element for eukaryotes—too much is toxic, but not enough is fatal.

*Tungsten:* a silvery-white lustrous metal. Tungsten has the highest melting point of all the metals, and also the highest tensile strength, meaning it can be used to form very strong wires. One of its more common applications is the filaments in incandescent lightbulbs. Tungsten has also been recently found to have a biological role in the unusual hyperthermophilic bacteria found near deep sea hydrothermal vents.

## References

1. Döbereiner JW. An Attempt to Group Elementary Substances According to Their Analogies. *Annalen der Physik und Chemie.* 1829; 15: 301–307.
2. Newlands JAR. On the Law of Octaves. *Chemical News.* 1865; 12: 83.
3. Moseley HGJ. The High Frequency Spectra of the Elements. *Phil. Mag.* 1913: 1024.
4. Emsley J. (1990). *The Elements.* Oxford University Press, New York.

## Procedure

### DETERMINE BULK DENSITIES OF THREE ELEMENTS

1. Weigh three samples of chromium, each weighing ~9 g; record the mass of each sample.
2. Weigh three samples of molybdenum, each weighing ~13 g; record the mass of each sample.
3. Weigh three samples of tungsten, each weighing ~20 g; record the mass of each sample.
4. Fill a graduated cylinder ~half full with water, and record the exact volume of water added.
5. Carefully add one sample of chromium using the forceps or tongs, being sure to not lose any of the water in the process (or break the cylinder!); record the new volume.
6. Repeat the process to add the other two chromium samples to the same container; again record the new volume after each addition.
7. Collect the metal pieces in a paper towel to dry; discard the water.
8. Repeat steps 4 through 7 for the molybdenum samples.
9. Repeat steps 4 through 7 for the tungsten samples.

## Data Collection

### DETERMINE BULK DENSITIES OF THE ELEMENTS

1. Chromium samples:

   Mass sample 1: _____

   Mass sample 2: _____

   Mass sample 3: _____

2. Molybdenum samples:

   Mass sample 1: _____

   Mass sample 2: _____

   Mass sample 3: _____

3. Tungsten samples:

   Mass sample 1: _____

   Mass sample 2: _____

   Mass sample 3: _____

4. Water volumes with chromium:

   Initial water volume: _____

   Total volume after addition of chromium sample 1: _____

Total volume after addition of chromium sample 2: _____

Total volume after addition of chromium sample 3: _____

5. Water volumes with molydenum:

Initial water volume: _____

Total volume after addition of molydenum sample 1: _____

Total volume after addition of molydenum sample 2: _____

Total volume after addition of molydenum sample 3: _____

6. Water volumes with tungsten:

Initial water volume: _____

Total volume after addition of tungsten sample 1: _____

Total volume after addition of tungsten sample 2: _____

Total volume after addition of tungsten sample 3: _____

## Analysis

### I. Determine Atomic Densities of the Elements

1. Determine the volume of one atom of each element (in units of cm$^3$ or mL):

$$\text{volume} = \frac{4\pi(\text{radius}_{\text{atom}})^3}{3}$$ **Equation 1**

Volume of chromium atom: _____

Volume of molybdenum atom: _____

Volume of tungsten atom: _____

2. Determine the mass of one atom of each element (in units of grams):

| | Include electrons | Do not include electrons |
|---|---|---|
| Mass of chromium atom: | _____ | _____ |
| Mass of molybdenum atom: | _____ | _____ |
| Mass of tungsten atom: | _____ | _____ |

3. Determine the density of one atom of each element using the data in Table 1 below (in units of g mL$^{-1}$)

$$\text{density} = \frac{[\text{mass(grams)}]}{[\text{volume(mL)}]}$$ **Equation 2**

| | Include electrons | Do not include electrons |
|---|---|---|
| Density of chromium atom: | _____ | _____ |
| Density of molybdenum atom: | _____ | _____ |
| Density of tungsten atom: | _____ | _____ |

## Table 1   Useful Data.

**Mass of Subatomic Particles (g)**

| | |
|---|---|
| proton | $1.672623 \times 10^{-24}$ |
| neutron | $1.674928 \times 10^{-24}$ |
| electron | $9.109387 \times 10^{-28}$ |

**Mass Number of Most Abundant Isotope**

| | |
|---|---|
| Chromium | 52 |
| Molybdenum | 98 |
| Tungsten | 184 |

**Atomic Radius (pm)**

| | |
|---|---|
| Chromium | 124.9 |
| Molybdenum | 136.2 |
| Tungsten | 137.0 |

## II.   Determine Bulk Densities of Three Elements

1.  Determine the volume of each chromium sample:

$$\text{volume} = (\text{measured volume})_{\text{final}} - (\text{measured volume})_{\text{initial}} \qquad \textbf{Equation 3}$$

Volume of chromium sample 1: _____

Volume of chromium sample 2: _____

Volume of chromium sample 3: _____

2.  Determine the volume of each molybdenum sample:

Volume of molybdenum sample 1: _____

Volume of molybdenum sample 2: _____

Volume of molybdenum sample 3: _____

3.  Determine the volume of each tungsten sample:

Volume of tungsten sample 1: _____

Volume of tungsten sample 2: _____

Volume of tungsten sample 3: _____

4.  Determine the bulk density of chromium:

Density of chromium sample 1: _____

Density of chromium sample 2: _____

Density of chromium sample 3: _____

Average density of chromium: _____

**5.** Determine the bulk density of molybdenum:

Density of molybdenum sample 1: _____

Density of molybdenum sample 2: _____

Density of molybdenum sample 3: _____

Average density of molybdenum: _____

**6.** Determine bulk density of tungsten:

Density of tungsten sample 1: _____

Density of tungsten sample 2: _____

Density of tungsten sample 3: _____

Average density of tungsten: _____

**7.** Percent error calculation (use literature values in Table 2 below).

$$\% \text{ error} = \frac{|(\text{density})_{\text{calculated}} - (\text{density})_{\text{literature value}}|}{(\text{density})_{\text{literature value}}} \times 100 \qquad \textbf{Equation 4}$$

Percent error for chromium: _____

Percent error for molybdenum: _____

Percent error for tungsten: _____

**Table 2   Literature Values for Densities.[4]**

| Element | Density (g cm$^{-3}$) |
|---|---|
| Chromium | 7.19 |
| Molybdenum | 10.2 |
| Tungsten | 19.3 |

## Reflection Questions

1.  Is there a trend in the atomic densities for your three elements as you progress down the family? If so, why would you expect that to be the case? If not, why should there not be a trend?

2.  Why did you make three experimental measurements of bulk density for each element?

3. What are possible reasons for differences in your bulk densities and bulk densities from the literature? Be specific.

4. As described earlier, bulk samples of metals (and other monatomic solids) form lattices that contain one of four repeating patterns (unit cells)—simple cubic (SC), body-centered cubic (BCC), hexagonal closest-packed (HCP), and cubic closest-packed (CCP) (Figure 6). As you consider the various packing structures, note the empty space between the atoms. So in a bulk solid, there are empty spaces that contain no mass (these spaces are smaller than the atoms around them)—the result is the density of a bulk solid sample is less than the atomic density of the atoms making up the solid. Each of the four packing arrangements has a different amount of filled space:

| | |
|---|---|
| Simple cubic (SC) | 52% of volume filled by atoms |
| Body-centered cubic (BCC) | 68% of volume filled by atoms |
| Hexagonal closest-packed (HCP) | 74% of volume filled by atoms |
| Cubic closest-packed (CCP) | 74% of volume filled by atoms |

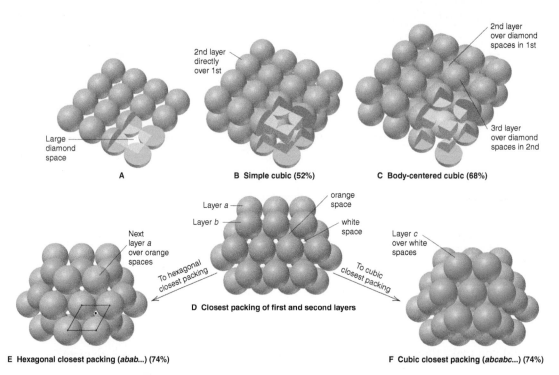

**Figure 6  Illustrations of different lattice (crystal packing) arrangements found in metals, as well as their packing efficiencies (% space filled by atoms shown in parentheses).**
(A) and (B) One and two layers, respectively, of a simple cubic (SC) structure. (C) BCC packing. (D) The first two layers of a closest packing arrangement. (E) and (F) Hexagonal closest packing (HCC) and cubic closest packing (CCP) arrangements, respectively.

(For example, 100% would indicate that all available volume is occupied by atoms, and 80% would indicate that 80% of the volume is occupied by atomic material, with 20% not occupied.)

For each of the metals that you have studied in this lab, you can calculate the percent of space occupied by atoms with the following calculation:

$$\% \text{ volume filled} = \frac{(\text{density})_{\text{bulk}}}{(\text{density})_{\text{atomic}}} \times 100 \qquad \textbf{Equation 5}$$

(a) Calculate the % volume filled for each element:

Chromium: _____

Molybdenum: _____

Tungsten: _____

(b) Based on your results in part (a), predict which of the four packing arrangements is used by each of these elements:

Chromium: _____

Molybdenum: _____

Tungsten: _____

5. Chromium is commonly found in a +3 charge state. When comparing Cr to $Cr^{3+}$, do the following properties increase, decrease, or remain unchanged?

Atomic/ionic size: _____

Ionization energy
(to remove outermost electron): _____

Electron affinity: _____

Atomic density: _____

6. The density of seaborgium, the man-made radioactive element at the bottom of the same family as chromium, molybdenum, and tungsten, is currently not known. On a graph, plot the row number of Cr, Mo, and W on the x-axis versus the average density of each element (determined

by you) on the *y*-axis. Using a ruler or straightedge, draw a "best-fit" straight line through the data points. Use this "best-fit" straight line to predict the density of seaborgium (assuming that Sg has the same crystal packing arrangement as the other three elements).

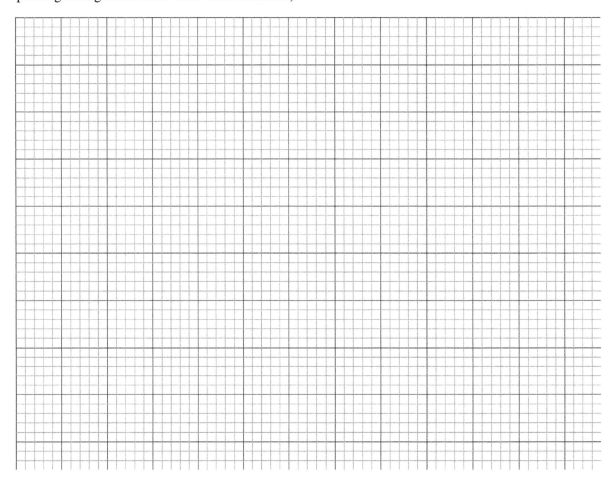

## Connection

Based on your experience in this lab, draw a connection to something in your everyday life or the world around you (something not mentioned in the background section):

# 6

# Molecular Geometry and Polarity

## Objectives

- Correlate Lewis structures with electron domain geometries.
- View the effect of lone pairs of electrons on molecular shape.
- Study the effect of polarity on the miscibility of liquids.
- Determine the type of solvent that generally dissolves ionic compounds.
- Determine the type of solvent that generally dissolves polar covalent compounds.
- Determine the type of solvent that generally dissolves nonpolar covalent compounds.
- Investigate the effect of adding a polar liquid solute to a nonpolar liquid solvent.

## Materials Needed

- 5 molecular models representing the 5 basic electron-domain geometries (prepared by instructor)
- 13 balloon models representing the 13 different molecular geometries (prepared by instructor)
- Test-tube rack
- Test tubes
- Stoppers to fit test tubes
- Pipettes
- Waste container
- Water
- Ethanol
- Cyclohexane
- Sodium chloride (table salt)
- Sucrose (table sugar)
- Naphthalene (moth balls)
- Unknown white solid

## Safety Precautions

- Standard safety precautions (goggles, gloves) are satisfactory for this lab.

## Background

The three-dimensional (3D) structure of a molecule (e.g., how its atoms are connected and arranged in space) is critical in determining its chemical and physical properties. As an example, a molecule's geometry determines whether or not it will be polar; polarity has a large impact on boiling and freezing points as well as chemical reactivity. The senses of smell and taste are largely governed by molecular structure. How drugs work and their specificity or toxicity are also controlled in part by geometry. It is therefore important to be able to predict the structure of molecules so that we can understand chemical and biological phenomena.

To determine the correct 3D structure for a molecule, the following strategy is used:

**Strategy:**

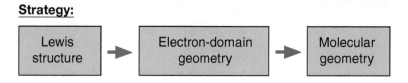

Step 1 – Draw the molecule's Lewis structure.

Step 2 – Determine the *electron-domain geometry* around the central atom.

Step 3 – Based on the number of lone pairs on the central atom, determine the *molecular geometry*.

Each of these steps is outlined next.

## STEP 1. Draw the Lewis structure.

A review of drawing Lewis structures is located at the end of this lab's workshop.

## STEP 2. Determine electron-domain geometry around the central atom.

Chemists utilize the **valence shell electron-pair repulsion (VSEPR) theory** to predict the 3D structure of molecules. VSEPR is a straightforward method that relates a molecule's Lewis structure (2D) to its molecular geometry, which can then be used to predict molecular properties such as polarity. VSEPR is based on the simple idea that electron ($e^-$) pairs (both bonding and nonbonding, or lone pairs) on an atom repel each other and try to stay as far apart from each other in space as possible. To minimize repulsion, $e^-$ pairs will distribute evenly around an atom, generating specific geometric arrangements, depending on the number of pairs.

We use the term **electron-domain geometry** to describe this geometric arrangement around an atom. An electron domain is composed of a single lone pair of electrons or a chemical bond; a single, double, or triple bond each count as a single electron domain. As just described, using $e^-$ pairs as an example, electron domains on a single atom will repel each other and therefore arrange to minimize electrostatic repulsions. There are five basic electron-domain geometries possible (Figure 1):

1. Linear: In this arrangement, there are only two electron domains on the middle atom. The two domains will orient themselves on opposite sides of the middle atom, generating a linear geometry with a bond angle of exactly 180°. Note that all two-atom molecules ($A_2$ or AB) will be linear molecules—there is no need to consider electron domains.

2. Trigonal Planar (or planar triangular): In this arrangement, all electron domains of the central atom are located in the same plane as the central atom, at the corners of a triangle (the central atom is located in the middle of the triangle). This, in turn, places all atoms in the molecule in the same plane, with bond angles of 120°. The three electron-domain positions are equivalent.

3. Tetrahedral: The four electron domains arrange at the corners of a pyramid, which contains four trigonal faces. The central atom is located in the center of the pyramid, with bond angles of 109.5°. As we saw for the trigonal planar electron-domain geometry, all four of the domain positions are equivalent.

4. Trigonal Bipyramidal: For this arrangement, there are five electron domains. As the name implies, this geometry can be thought of as containing two pyramids that share one trigonal face; the trigonal bipyramid has six trigonal faces, with one domain at each corner; and the central atom is located in the center of the equatorial triangle. This geometry contains two different types of positions: three equatorial positions (corners of shared trigonal face) and two axial positions (one above and one below the shared trigonal face). There are also two different bond angles: 120° between the equatorial positions and 90° between an axial position and each of the equatorial positions.

| Electron Domain Geometry (GEOMETRY) | Molecular Geometry (SHAPE) |
|---|---|

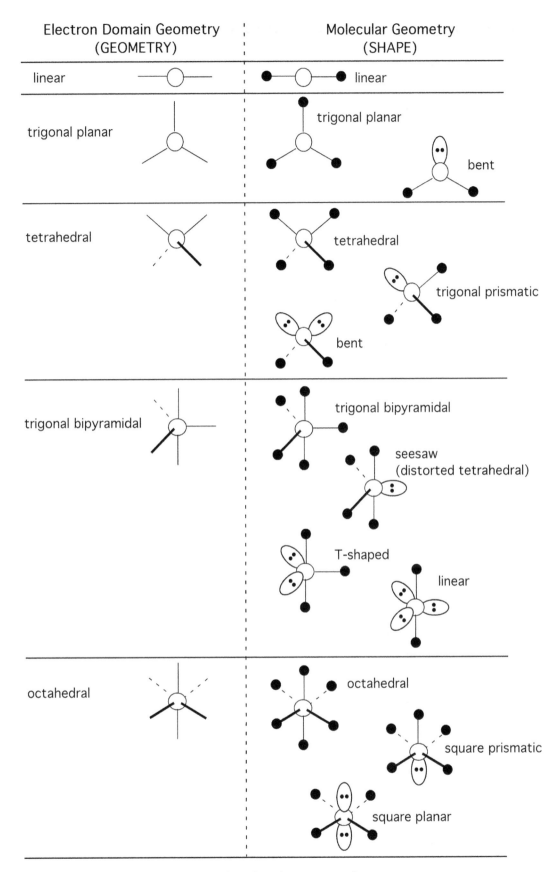

**Figure 1** **Electron-domain geometries and molecular geometries.**

5. Octahedral: Six electron domains arrange themselves at the corners of an octahedron, with the central atom in the center equidistant from the six corners. An octahedron can be thought of as two square pyramids that share their square bases (eight sides total). This creates 90° bond angles between all corners (all domain positions are equivalent).

## STEP 3. Determine the molecular geometry.

The *electron-domain geometry* describes where the electron domains are on the middle atom; the **molecular geometry** (or shape) of the molecule describes where the atoms (or nuclei) are in the molecule. The distinction comes from the fact that an electron domain may be involved in a bond between the middle atom and an outer atom, or else may simply be a lone pair on the middle atom and therefore will occupy space but will not be involved in a bond. A single electron-domain geometry can be the basis for multiple molecular geometries, depending on the number of lone pairs on the middle atom (Figure 1 and Table 1):

1. Trigonal planar electron-domain geometry:
   * Trigonal planar molecular geometry (no lone pairs on middle atom)
   * Bent molecular geometry (one lone pair on middle atom)

**Table 1   The VSEPR Structures for Molecules and Ions.**

| Valence Electron Pairs | Bonding Electron Pairs | Nonbonding Electron Pairs | Approx. Bond Angles | Electron-domain Geometry (geometry) | Molecular Geometry (shape) |
|---|---|---|---|---|---|
| 2 | 2 | 0 | 180° | Linear | Linear |
| 3 | 3 | 0 | 120° | Trigonal planar | Trigonal planar |
|  | 2 | 1 | < 120° | Trigonal planar | Bent |
| 4 | 4 | 0 | 109.5° | Tetrahedral | Tetrahedral |
|  | 3 | 1 | < 109.5° | Tetrahedral | Trigonal pyramidal |
|  | 2 | 2 | < 109.5° | Tetrahedral | Bent |
| 5 | 5 | 0 | 90° and 120° | Trigonal bipyramidal | Trigonal bipyramidal |
|  | 4 | 1 | < 90° and < 120° | Trigonal bipyramidal | Seesaw (irregular tetrahedral) |
|  | 3 | 2 | < 90° | Trigonal bipyramidal | T-shaped |
|  | 2 | 3 | 180° | Trigonal bipyramidal | Linear |
| 6 | 6 | 0 | 90° | Octahedral | Octahedral |
|  | 5 | 1 | < 90° | Octahedral | Square pyramidal |
|  | 4 | 2 | 90° | Octahedral | Square planar |

2. Tetrahedral electron-domain geometry:
   - Tetrahedral molecular geometry (no lone pairs on middle atom)
   - Trigonal prismatic molecular geometry (one lone pair)
   - Bent molecular geometry (two lone pairs)
3. Trigonal bipyramidal electron-domain geometry:
   - Trigonal bipyramidal molecular geometry (no lone pairs on middle atom)
   - Seesaw or distorted tetrahedral molecular geometry (one lone pair): the five positions are not equivalent; the lone pair goes into an equatorial position
   - T-shape molecular geometry (two lone pairs): the two lone pairs go into equatorial positions
   - Linear molecular geometry (three lone pairs): all three lone pairs go into equatorial positions
4. Octahedral electron-domain geometry:
   - Octahedral molecular geometry (no lone pairs on middle atom)
   - Square pyramidal molecular geometry (one lone pair): all positions are identical, so it does not matter where the single lone pair goes
   - Square planar molecular geometry (two lone pairs): the two lone pairs will locate opposite each other

## DEVIATION FROM IDEAL BOND ANGLES

Lone pairs of electrons occupy more space than bonding pairs of electrons (the bonded pairs are attracted to two nuclei, while a lone pair is only attracted to a single nucleus and thus spreads out). This leads to a distortion of the ideal bond angles in molecular geometries that contain one or more lone pairs on the middle atom. Similar distortions are generated when a double or triple bond is located at one of the electron domain positions; the generally accepted trend in terms of electron domain space requirements is shown in Figure 2.

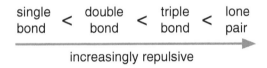

Figure 2  **Relative space requirements of electron domains.**

Figure 3 shows different molecular geometries with observed bond angles; the lone pair on the N in $NH_3$ pushes the three H atoms away, decreasing the H–N–H angles from 109.5° to 107°; the C = O double bond repels the H atoms, decreasing the H–C–H angle from 120° to 116°; similar distortions are generated by a single lone pair on S in $SF_4$ and on Br in $BrF_5$.

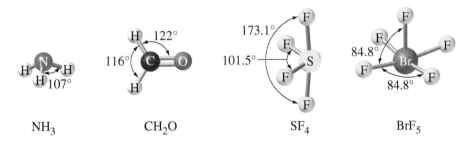

Figure 3  **Deviations from ideal bond angles in molecular geometries.**

# MOLECULAR GEOMETRY AND POLARITY

Polarity results from the molecular geometry (shape) of a molecule and the relative electronegativities of the component atoms. To determine whether a molecule is polar, one must first consider the polarity of each individual bond in the structure. If the molecule/ion is diatomic (like HCl), then the polarity of the molecule will be equivalent to the polarity of the single bond. With more than two atoms, we must look at the arrangement of the polar bonds to see if there is a net dipole for the structure. Two specific examples follow. The first is carbon dioxide ($CO_2$). This molecule contains two polar C = O bonds; the two bond dipoles point exactly opposite of each other, however, so the dipoles counter each other. The net result is no net molecular dipole—that is, $CO_2$ is not polar. The second example is sulfur dichloride ($SCl_2$). This molecule again contains two polar S–Cl bonds; due to lone pairs of electrons on the S atom, the molecule adopts a bent shape, so the S–Cl bonds are not pointed directly opposite one another, so they do not cancel each other completely. The net result is an overall dipole; thus, $SCl_2$ is a polar molecule. In general, a molecule will be *nonpolar* if (1) there are no lone pairs of electrons on the middle atom and (2) all of the bonds have similar dipole moments and so cancel out. A molecule has a good chance of being *polar* if the middle atom contains one or more lone pairs, and there is an asymmetric arrangement of peripheral atoms. The polarity of the molecule ultimately depends, however, on the arrangement of the polar bonds around the middle atom (i.e., molecular geometry) and the electronegativity differences between the bonded atoms.

Polarity affects the boiling point, melting point, and many other properties of a molecule. Polarity also plays an important role in determining if two substances can combine to form a homogenous mixture (called a solution). Homogeneous means uniform throughout. When a solution forms, the attractive interactions between the particles (ions, molecules, atoms) of each substance must be disrupted and replaced with interactions involving the other substance in the mixture. Those attractive interactions are based largely on the polarity of each substance. A general rule is that "like dissolves like," indicating that substances with similar polarities will mix favorably and form a solution, while two substances with very different polarities will usually not mix effectively. People who work in the dry-cleaning industry, for example, utilize solvents to remove stains from clothing. In many cases, the nature of the stain will determine the polarity of the dry-cleaning solvent that is used.

## Procedure/Data Collection

Stations: Groups of students will rotate between four stations. Groups will spend 20 to 25 minutes at each station. Within a group, students are encouraged to work together, but each student must fill in (and turn in) a separate worksheet for this lab. Worksheets are turned in at the end of the lab today.

## Station 1. Lewis Structures and Molecular Models

In your workshop, you were assigned one of the sets of eight substances (shown in Table 2) for which you drew Lewis structures.

| Table 2 | Sets of Substances. | | | | | | | |
|---------|---------|---------|---------|---------|---------|---------|---------|---------|
| **Set** | **Substances** | | | | | | | |
| A | $CF_3Cl$ | $ClO_2^-$ | $AsF_6^-$ | $SeBr_3^+$ | $I_3^-$ | $O_3$ | $ICl_4^+$ | $XeBr_4$ |
| B | $SnCl_4$ | $BrO_2^-$ | $AlF_6^{3-}$ | $H_3O^+$ | $XeF_2$ | $AsO_2^-$ | $SF_4$ | $ICl_4^-$ |
| C | $ClO_4^-$ | $OI_2$ | $SbF_6^-$ | $PF_3$ | $Br_3^-$ | $NO_2^-$ | $TeCl_4$ | $KrF_4$ |
| D | $ClF_2O_2^+$ | $SCl_2$ | $BrF_6^+$ | $SeI_3^+$ | $ICl_2^-$ | $SO_2$ | $SeBr_4$ | $ClF_4^-$ |
| E | $CFCl_3$ | $ClF_2^+$ | $SeF_6$ | $SBr_3^+$ | $KrF_2$ | $SeO_2$ | $ClF_4^+$ | $XeI_4$ |

For each of the eight Lewis structures from your workshop (Table 2), you are to match the structure to one of the five molecular models on the lab bench, each of which represents one of the possible *electron-domain geometries*. In Table 3, for each,

- Identify the correct model number.
- Name the electron-domain geometry (geometry).
- Draw the 3D structure for your Lewis structure (do not indicate potential distortions to the ideal geometry)—make it clear!!
- Label all ideal bond angles in your structure.

| Table 3   Electron-domain Geometries. | | | |
|---|---|---|---|
| **Formula** | **Model** | **Electron-domain Geometry Name** | **Drawing with Angles Labeled** |
| | | | |
| | | | |
| | | | |
| | | | |

*(continued)*

Table 3   Electron-domain Geometries—(*continued*)

| Formula | Model | Electron-domain Geometry Name | Drawing with Angles Labeled |
|---------|-------|-------------------------------|------------------------------|
|         |       |                               |                              |
|         |       |                               |                              |
|         |       |                               |                              |
|         |       |                               |                              |

## Station 2. VSEPR and Balloon Models: effect of lone pairs on molecular shape

For each of the eight Lewis structures from your workshop (Table 2), you are to match the structure to one of the balloon models that represent various VSEPR structures. The balloons represent the space occupied by bonding pairs or lone pairs. The central atom resides at the central knot where the balloons are joined. In Table 4, for each of your structures from your workshop,

- Identify the correct VSEPR model number.
- Name the molecular geometry (shape).
- Draw the 3D structure for your Lewis structure, clearly indicating expected distortions.
- Label all bond angles in your structure, indicating distortions using "<" or ">" signs.

## Table 4   Molecular Geometries.

| Formula | Model | Molecular Geometry Name | Drawing with Angles Labeled |
|---------|-------|-------------------------|------------------------------|
|         |       |                         |                              |
|         |       |                         |                              |
|         |       |                         |                              |
|         |       |                         |                              |
|         |       |                         |                              |

*(continued)*

Table 4   Molecular Geometries—(*continued*)

| Formula | Model | Molecular Geometry Name | Drawing with Angles Labeled |
|---------|-------|-------------------------|-----------------------------|
|         |       |                         |                             |
|         |       |                         |                             |
|         |       |                         |                             |

## Station 3. Polarity of Solvents: mixing of solvents of different polarities

1. Add one full pipette of water to test tubes 1 and 2.
2. Add one full pipette of ethanol to test tubes 1 and 3.
3. Add one full pipette of cyclohexane to test tubes 2 and 3.
4. Stopper the open end of each test tube and agitate the liquids to thoroughly mix.
5. Examine what happens to the liquids after agitation and record your observations in Table 5.
6. Dispose of liquids properly and clean your glassware.

| Table 5   Mixing Liquids. | |
|---------------------------|---|
| **Solvent Mixture Tested** | **Results Observed** |
| Water/ethanol | |
| Water/cyclohexane | |
| Ethanol/cyclohexane | |

## Station 4. "Like Dissolves Like": determination of polarity of unknown solid

1. Add one full pipette of water into test tubes 1 through 4.
2. Add one full pipette of ethanol into test tubes 5 through 8.
3. Add one full pipette of cyclohexane into test tubes 9 through 12; you should now have a 3 × 4 grid of solvent-filled tubes.
4. In tubes 1, 5, and 9 add enough NaCl to cover the bottom of the tube. Stopper those tubes and invert to agitate the mixture. Repeat several times until no further changes take place. Carefully examine the contents of each tube and record your observations in Table 6.
5. Repeat step 4 with sucrose (tubes 2, 6, and 10).
6. Repeat step 4 with naphthalene (tubes 3, 7, and 11).
7. Repeat step 4 with the unknown solid (tubes 4, 8, and 12).
8. Dispose of tube contents properly and clean your glassware.

| Table 6    Mixing Solids and Liquids. | | | |
|---|---|---|---|
| | Solvents | | |
| Solutes | Water | Ethanol | Cyclohexane |
| NaCl | | | |
| Sucrose | | | |
| Naphthalene | | | |
| Unknown | | | |

## Reflection Questions

1. For each solvent, draw its Lewis structure (the skeletons for ethanol and cyclohexane are given).

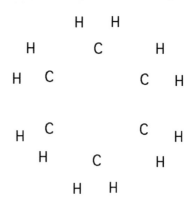

```
     H    H
 H   C    C   O   H
     H    H
```

2. Based on the Lewis structures in question 1, determine the type for each (polar or nonpolar).

Water _____

Ethanol _____

Cyclohexane _____

3. Based on the Lewis structures you drew in your workshop (Table 2), predict whether each compound is polar or nonpolar. [Note: A polyatomic ion can also be considered polar or nonpolar, depending on the arrangement of polar bonds in the structure. An ion is charged, meaning it has a different number of electrons versus protons; something is polar, on the other hand, if its electron density is unevenly distributed over its structure. Soluble ionic substances will dissolve in polar solvents and not in nonpolar solvents (whether or not the individual ions are polar) because the individual ions are charged.]

| Formula | Polar or Nonpolar? |
|---------|--------------------|
|         |                    |
|         |                    |
|         |                    |
|         |                    |
|         |                    |
|         |                    |
|         |                    |
|         |                    |

4. What general trend appears in Table 6 with regard to which type of solute dissolves in which type of solvent?

5. Classify the known solids in Table 6 as ionic, polar covalent, or nonpolar covalent.

6. Attempt to classify the unknown solid as ionic, polar covalent, or nonpolar covalent.

7. Explain which solvent from this experiment you would use to remove road salt stains from a pair of jeans (road salts are chloride-based salts added to roads in the northeast to combat icing).

8. What general rule can be followed when choosing a type of solvent to dissolve a particular solid?

## Connection

Based on your experience in this lab, draw a connection to something in your everyday life or the world around you (something not mentioned in the background section).

# 7

# Percent Composition from Gravimetric Analysis: Calcium Carbonate in Texas Limestone

## Objectives

- Use an understanding of solubility rules and precipitation reactions to isolate calcium (and ions, in general) from a real-world sample.
- Use mass relationships to determine the percent composition of calcium carbonate in limestone.

## Materials Needed

- ~4.0 g of limestone (two ~1.5- to 2.0-g samples)
- 3.00 $M$ HCl (50 mL)
- 125-mL Erlenmeyer flask (2)
- Gooch crucible (2)
- Filter paper (e.g., Fisherbrand grade P5)
- Methyl red indicator
- 6.00 g ammonium sulfate [$(NH_4)_2SO_4$]
- 2.00 $M$ aqueous ammonia ($NH_3$) solution [also sometimes labeled ammonium hydroxide ($NH_4OH$)] (~50 mL)
- 250-mL Erlenmeyer flask

## Safety Precautions

- Gloves and safety goggles/glasses should be worn during this lab.
- Hydrochloric acid (HCl) is toxic by ingestion and inhalation and corrosive to skin and eyes; avoid contact with body tissues.
- Aqueous ammonia ($NH_3$) solution is corrosive, and contact with the skin or eyes should be avoided. Work in a well-ventilated area.

## Background

Chemical structure leads to observable properties: drug discovery centers around the determination of chemical structures that display biological activity; and the development of new polymers, materials, dyes, electronics, alternative energy sources, etc. are all fundamentally linked to an understanding of *structure-property* relationships. It is simply impossible to avoid the importance of structure in the description of properties and applications for essentially anything! From the discovery of fire (wood = structure; burns = property), inquiries into *structure-property* relationships have served as the foundation to explain the world around us while also allowing for the prediction of future possibilities.

An understanding of the properties of a substance begins with the determination of its elemental composition. From that we can determine a chemical formula and then a bonded arrangement for that composition, which ultimately guides our understanding of its properties. Percent composition (or percent by mass) is one way to express the elemental makeup of a pure substance. It can be determined from an empirical or molecular formula:

$$\% \text{ composition} = \frac{(\text{mass of element})}{(\text{molar mass of compound})} \times 100 \qquad \textbf{Equation 1}$$

This is also known as the theoretical percent composition. In a similar way, the percent composition can also be expressed for an impure substance:

$$\% \text{ composition} = \frac{(\text{mass of component of interest})}{(\text{total mass of sample})} \times 100 \qquad \textbf{Equation 2}$$

To determine the percent composition of a sample, one needs to identify what elements are present, and in what quantities. One common way to accomplish this is to add a compound (precipitating agent) to a dissolved sample that combines selectively with the element of interest (the analyte), causing it to form a solid precipitate of known identity and composition. The precipitate is then isolated and weighed to quantify the analyte. This method is known as *gravimetric analysis*. Gravimetric analysis is used in real-world applications, such as in the determination of the composition of rocks, minerals, and alloys, and in the quality control of raw materials and finished products in industry.

In this lab, we will be using gravimetric analysis to determine the amount of calcium carbonate ($CaCO_3$) in a sample of limestone. Calcium salts (especially $CaCO_3$) lend strength to a number of naturally occurring objects, including rocks, sea shells, pearls, and egg shells. $CaCO_3$ is unusual in that its solubility (i.e., how much $CaCO_3$ will be dissolved in a given amount of water) increases with decreasing temperature and is also impacted by pressure ($CO_2$ partial pressure, specifically), properties not usually identified with many solids.

Limestone is a type of rock that contains large amounts of $CaCO_3$ (in the form of calcite or aragonite, which differ in their crystalline forms). Limestone comprises approximately 10% of all sedimentary rocks exposed on the surface of the earth (Figure 1) and is primarily of biological origin. These rocks typically form through lithification of accumulated shells, corals, algae and other skeletal debris on the ocean floor. Less commonly, limestone can be formed by the direct precipitation of $CaCO_3$ from water (e.g., the combination of $CO_2$, $Ca^{2+}$ and water). In caves, beautiful flowstone and limestone stalactites and stalagmites form from the slow evaporation of saturated solutions of calcium carbonate. The purest form of limestone, called calcite (Figure 2), is essentially pure $CaCO_3$ (96 to 99% $CaCO_3$) and is the main component of natural chalk. Chalk (Figure 2) was used in prehistoric times by early man for cave drawings, and has been a primary part of common chalk used in schools [although today many types of pastel chalks are composed not of $CaCO_3$, but instead of a related salt, calcium sulfate ($CaSO_4$),

**Figure 1   The White Cliffs of Dover (Kent, England).** At over 350 ft high, these 136-million-year-old formations are composed of chalk (pure white lime) and mark the location where Great Britain most closely approaches continental Europe.

derived from gypsum]. Other forms of limestone contain various amounts of Mg, silicates, Mn, Fe, Ti, Al, Na, K, S (in the form of sulfides, sulfates, or oxides) and P; these are mineral contaminants that come mainly from the seawater in which the limestone was originally created. As calcium carbonate is a white substance, discoloration in a limestone sample is due to the presence of contaminants. Limestone that has undergone a natural recrystallization process caused by heat and pressure in the Earth's crust becomes marble.

**Figure 2   Samples of calcite (left) and chalk (right).**

Limestone is the most common sedimentary rock in Texas, and has been used for thousands of years for numerous applications. Over 9000 years ago, prehistoric people in the Texas Plateaus and Canyonlands (the hill country in central Texas) used limestone rocks as cooking stones; large chunks of limestone can be heated to close to 500°C in a simple wood fire, and will hold this heat for hours for use in primitive ovens. Texas "holey rock" or "honeycomb limestone," endemic to Texas, contains numerous holes that come from the action of acidic groundwater moving through cracks in the stone and etching away the rock (often leading to underground caves or caverns, see Figure 3). Other uses of limestone include:

- Construction materials (road base, concrete aggregate, in cement, mortar)
- Building stone (commonly used in the nineteenth and twentieth centuries in libraries, office buildings, banks, etc.; also found in many well-known monuments around the world including the Great Pyramid in Egypt!)
- Quicklime (added to acidic soils to neutralize pH)

**Figure 3   Carlsbad Caverns in New Mexico.**

- Animal feed filler (Ca for chickens and dairy cattle)
- Nutritional supplements and personal care products (Ca pills, toothpastes)
- Water purification and sewage treatment (for limestone with high amounts of Ca)

In this lab, you will be determining the amount of $CaCO_3$ in a sample of limestone using two different gravimetric analysis methods. As calcium carbonate is basic, it will react with acids. In the first step of this lab, you will dissolve a limestone sample with a strong acid (3.0 $M$ HCl), resulting in the initial formation of carbonic acid, $H_2CO_3$, which decomposes to give $CO_2(g)$ as described below:

$$CaCO_3(s) + 2HCl(aq) \rightarrow CaCl_2(aq) + H_2CO_3(aq) \qquad \textbf{Reaction 1}$$

$$H_2CO_3(aq) \rightarrow CO_2(g) + H_2O(l) \qquad \textbf{Reaction 2}$$

*net reaction:* $\quad CaCO_3(s) + 2HCl(aq) \rightarrow CaCl_2(aq) + CO_2(g) + H_2O(l) \qquad \textbf{Reaction 3}$

By measuring the change in mass of the limestone/acid mixture (i.e., before and after $CO_2$ gas is released), one can determine the amount of $CaCO_3$ in the original sample (based on reactions 1 and 2, each $CO_2$ molecule comes from one carbonate ion, and thus one $CaCO_3$ in the original sample—this is called a *stoichiometric* relationship). *Thus, the $CaCO_3$ is quantified by analyzing the anion of the salt.*

We will also quantify the $CaCO_3$ by precipitating the cation $Ca^{2+}$ with the precipitating agent sulfate ($SO_4^{2-}$), forming $CaSO_4$ (which is gypsum, the mineral currently used in the commercial production of chalk):

$$CaCl_2(aq) + (NH_4)_2SO_4(aq) \rightarrow CaSO_4(s) + 2NH_4Cl(aq) \quad \textbf{Reaction 4}$$

*By measuring the mass of the $CaSO_4$ sample, we can determine the amount of Ca (and thus $CaCO_3$) in the original limestone sample.* A comparison of the % compositions that we obtain from the cation and anion analyses will allow for comparative results.

## Procedure

### I. Carbonate in the Limestone

1. Using a hammer (or other heavy object), break a piece of limestone into small pieces—ideally, the pieces should have masses no greater than 1 g each.
2. Select one (or a few) pieces of limestone that have a total mass of ~1.5 to 2.0 g; record the mass.
3. Add approximately 25 mL of 3.00 $M$ HCl to a clean and dry (important!) 125-mL Erlenmeyer flask. Record the total mass of the container with the HCl.
4. Add the limestone to the flask with the HCl—the limestone should begin to dissolve with the formation of bubbles (production of $CO_2$ gas). Allow the reaction to proceed until the mixture stops bubbling/foaming, indicating that the reaction is complete.
5. Record the final mass of the container + remaining contents. *Do not dispose of the solution*—you will continue using this mixture for Part II.
6. Repeat steps 1 to 5. This will give you two sets of data for Part I, and two samples to move forward to Part II.

### II. Calcium in the Limestone

For each sample from Part I:

7. The limestone sample in Part I should have completely dissolved in the HCl solution. If any solid did not dissolve (impurities such as silicates, sulfides, and sulfates), one can carefully filter at this step and place the filtrate into a clean 125-mL Erlenmeyer flask for use in the rest of the lab.
8. Measure the mass of a clean and dry Gooch crucible together with a piece of filter paper placed inside the crucible.

9. Add 10 drops of a methyl red indicator to either the reaction mixture from Part I (if you did not need to filter) or filtrate from step 7; methyl red is used to indicate when changes in pH occur (where pH is a measure of acidity)—the solution should turn pink (indicating a pH < 4.4).

10. Add 3.00 g of $(NH_4)_2SO_4$ to the flask while stirring. The $(NH_4)_2SO_4$ should completely dissolve as a cream-colored precipitate appears.

11. Then, with constant stirring, slowly add 2.00 $M$ $NH_3$ until the color *just* shifts from pink to a pale yellow (this indicates a change in pH to approximately 6.2). It is important not to overshoot this color change significantly (i.e., do not keep adding $NH_3$ once the color changes and is maintained with stirring or shaking).

12. Filter the solution/precipitate mixture through the Gooch crucible/filter paper into a 250-mL Erlenmeyer flask; then wash any remaining solid from the flask with cold water.

13. Once the filtrate has completely passed through the crucible, wash with water and then transfer the crucible (with precipitate) to an oven to dry overnight or set aside to air dry.

14. Once dry, record the mass of the crucible/filter paper plus the precipitate.

## Data Collection

### I.  Carbonate in the Limestone

1. Mass of limestone:

    Sample 1: _____

    Sample 2: _____

2. Mass of container + HCl:

    Sample 1: _____

    Sample 2: _____

3. Mass of additional 5 mL HCl (if added):

    Sample 1: _____

    Sample 2: _____

4. Final mass of container + contents:

    Sample 1: _____

    Sample 2: _____

### II.  Calcium in the Limestone

5. Mass of Gooch crucible/filter paper:

    Sample 1: _____

    Sample 2: _____

6. Mass of crucible/filter paper + ppt:

    Sample 1: _____

    Sample 2: _____

## Analysis

### I.  Carbonate in the Limestone

7. Mass of $CO_2$ released = [(#1) + (#2) + (#3)] − (#4)

    Sample 1: _____

    Sample 2: _____

8. Moles of $CO_2$ released

    Sample 1: _____

    Sample 2: _____

9. Moles of carbonate in limestone

    Sample 1: _____

    Sample 2: _____

**10.** % $CaCO_3$ in limestone (Equation 2)

Sample 1: _____

Sample 2: _____

Average: _____

## II. Calcium in the Limestone

**11.** Mass of $CaSO_4$ ppt = (#6) − (#5)

Sample 1: _____

Sample 2: _____

**12.** Moles $CaSO_4$ ppt

Sample 1: _____

Sample 2: _____

**13.** Moles $CaCO_3$ in limestone

Sample 1: _____

Sample 2: _____

**14.** Mass of $CaCO_3$ in limestone

Sample 1: _____

Sample 2: _____

**15.** % $CaCO_3$ in limestone (Equation 2)

Sample 1: _____

Sample 2: _____

Average: _____

## Reflection Questions

1. How do you know that sufficient acid was added to completely react with all of the calcium carbonate? If you failed to add sufficient HCl, how would this affect your percent composition calculations?

2. (a) In Part I (Carbonate in Limestone), what are the possible sources of error?

   (b) In Part II (Calcium in Limestone), what are the possible sources of error?

3. Suppose you were to take the solution from Part I and remove all solvent. Could you then determine the difference between the mass of this sample and the initial mass of the limestone to obtain the amount of $CO_2$ lost? If the answer is no, then what is flawed with this approach?

4. If you wanted to be really careful, how would you check to see if your calcium sulfate is completely dry?

5. Compare your % CaCO₃ from Parts I and II. If they are similar, what does that tell you about the calcium and carbonate in your limestone sample? If they are different, what could be the reason why that would be true?

6. Acid rain (more generally known as acid precipitation) is an environmental problem we face today that comes from the reaction of SO$_x$ and NO$_x$ compounds in the environment with moisture, yielding acids such as HNO₃ and H₂SO₄. The SO$_x$ and NO$_x$ comes primarily from human activities, namely, the burning of sulfur-containing coal and operation of motor vehicles, as well as various natural sources such as lightning and volcanic activity. The production of these acids leads to environmental damage from acidification of bodies of water, which hurts forests and other vegetation, insects, and numerous forms of aquatic life. We also observe damage to marble features on buildings and statuary.

Based on the chemistry you observed in this lab, write the two chemical reactions for the reaction of these two acids with marble:

7. In Part I of this lab, you measured the mass of the limestone sample that was lost as carbon dioxide. The procedure and your analysis assumed that all of the carbon dioxide gas escaped into the atmosphere. However, carbon dioxide has some solubility in water. In fact, at room temperature and atmospheric pressure, the solubility of carbon dioxide in water is $1.45$ g $L^{-1}$.

(a) Assuming a similar solubility in your acidic solution, how much carbon dioxide in grams remained in your solution?

(b) In words, how would this affect your carbonate analysis?

(c) Correct your final carbonate analysis to reflect all $CO_2$ lost (both escaped in the gas phase and in solution).

## Connection

Based on your experience in this lab, draw a connection to something in your everyday life or the world around you (something not mentioned in the background section):

# 8

# Limiting Reactant Lab

## Objectives

- Understand mole-to-mole ratios in a balanced chemical reaction.
- Identify limiting and excess reactants in a chemical reaction based on experimental observations.
- Conduct limiting reactant calculations to predict theoretical yields.

## Materials Needed

- 250-mL beakers (three)
- 250-mL Erlenmeyer flask
- Glass stirring rod(s)
- 50- or 100-mL graduated cylinder
- Copper(II) chloride dihydrate [$CuCl_2 \cdot 2\, H_2O$] (three samples: 0.50 g, 0.70 g, and 0.70 g)
- Aluminum foil [$Al(s)$] (three samples: 0.25 g, 0.05 g, and to-be-determined—less than 0.25 g)
- Deionized water
- Methanol
- 6 $M$ Hydrochloric acid (HCl) solution (prepared by instructor)

## Safety Precautions

- Gloves and safety goggles/glasses should be worn during this lab
- NaOH is a strong base that can cause burns. Avoid contact with body tissues.
- Methanol is toxic if ingested or inhaled. It is also very flammable. Do not use methanol near open flames.
- Concentrated hydrochloric acid is a strong acid and is toxic by ingestion and inhalation and corrosive to skin and eyes; avoid contact with body tissues. It must be stored and used in a laboratory hood.
- Dispose of waste solutions in designated waste containers as indicated by your instructor.

## Background

Following a recipe is an important skill in life—from baking the perfect cake, to preparing the right amount of cement for that new sidewalk you are installing, to putting that toy together for your 6-year-old boy on Christmas morning. All of these involve taking a recipe or set of instructions and a pile of resources, and from those generating a specific outcome. In chemistry, a balanced chemical equation is the recipe that describes a specific chemical reaction and defines exact atom-to-atom (or molecule-to-molecule, or mole-to-mole) ratios; it can be used to determine exactly how much of each product can theoretically be made. To determine the actual yield of a reaction or how much of each product is generated, one needs to know (1) the amounts of starting materials (reactants) and (2) the percent yield of the reaction, a measure of the efficiency of the reaction.

Experimental conditions, for example, temperature, pressure, choice of solvent, and the like, impact reaction outcomes and, thus, can be adjusted to increase the yield of a desired product of a chemical reaction.

## CALCULATIONS WITH BALANCED CHEMICAL EQUATIONS

A *balanced* chemical equation allows us to determine quantitatively how substances react. Substances (atoms, molecules, or compounds) react according to fixed molar ratios, so only a limited amount of product can form from given amounts of starting materials. These fixed molar ratios are defined as the *stoichiometry* of the reaction. Consider, for example, chemical equation 1:

$$3NaOH + H_3PO_4 \rightarrow Na_3PO_4 + 3H_2O \qquad \textbf{Equation 1}$$

The coefficients in the equation (i.e., the numbers that precede each compound formula) define the stoichiometric ratios in the reaction (if no number is written before the formula, the coefficient is "1"). The balanced equation reflects that matter is conserved during a chemical reaction (a requirement!). Chemical equation 1 defines a number of stoichiometric ratios:

- Every 3 mol (or formula units) of NaOH will react with exactly 1 mol (or molecule) of $H_3PO_4$.
- Every 3 mol (or formula units) of NaOH will yield exactly 1 mol (or formula unit) of $Na_3PO_4$ and 3 mol (or molecules) of $H_2O$.

These molar ratios can be expressed as fractions or conversion factors:

$$\frac{3 \text{ mol NaOH}}{1 \text{ mol } H_3PO_4} \qquad \frac{3 \text{ mol NaOH}}{1 \text{ mol } Na_3PO_4} \qquad \frac{3 \text{ mol NaOH}}{3 \text{ mol } H_2O}$$

*IMPORTANT POINT 1: A balanced chemical equation defines exact MOLAR (or molecular, or formula, or atomic) ratios.*

So, "3 mol of NaOH react with exactly 1 mol of $H_3PO_4$" is *correct*, but "3 g of NaOH react with exactly 1 g of $H_3PO_4$" is *not correct* based on Equation 1. When using a balanced chemical equation to do a stoichiometry calculation, one will be working in units of moles (or number of atoms or molecules). If given a different starting unit, one must first convert to moles, as illustrated in Figure 1:

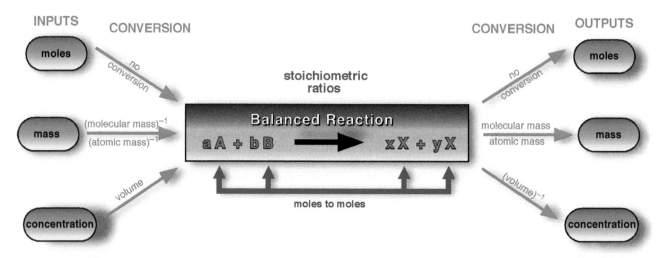

**Figure 1  Schematic showing the relationship of reactant and product amounts and their units to a balanced equation—a practical guide to performing stoichiometric calculations.** The "inputs" are the common ways in which we measure amounts of reactants, each with their characteristic units. The balanced equation defines *molar* ratios, and so any starting units must be converted to moles. Once the stoichiometric calculation is finished, the answer can be converted to a different "output" unit, depending on what is asked for in the problem. *All stoichiometry problems, therefore, can be reduced to essentially a single problem or can be solved by a general method*: (1) convert the input data to moles; (2) use the balanced chemical equation to define the appropriate molar ratios and determine the moles of the desired compound; (3) finally, convert to the desired output units.

Molar ratios can then be used as conversion factors to determine what the outcome will be from a specified amount of reactants. For example, if one mixes 5 mol NaOH with an excess of $H_3PO_4$, how much $Na_3PO_4$ will be produced? Following is the calculation to solve this:

$$(5 \text{ mol NaOH}) \times \left( \frac{1 \text{ mol } Na_3PO_4}{3 \text{ mol NaOH}} \right) = {}^5\!/_3 \text{ mol } Na_3PO_4 \text{ produced}$$

The conversion factor (molar ratio from the balanced equation) is written with the "mol $Na_3PO_4$" in the numerator and "mol NaOH" in the denominator so that the "mol NaOH" units cancel out, leaving an answer in units of "mol $Na_3PO_4$."

> *IMPORTANT POINT 2: Identification and proper use of conversion factors helps ensure that the answer to a stoichiometry problem will be correct and reported with the proper units.*

## LIMITING REACTANT CALCULATIONS

It is often the case when performing a chemical reaction that the exact stoichiometrically correct amount of every reactant will *not* be present. When that is the case, there will be one compound that limits the amount of product that can be generated—this is the *limiting reactant*; the other reactant(s) will be present in abundance (more than is needed) and is/are called the *excess reactant(s)*. Once the reaction is complete, the limiting reactant is completely consumed, while the excess other reactants remain.

To better understand the concept of the limiting reactant, let's bake some apple pies! Suppose that based on the recipe, a single apple pie requires 1 pie crust and 4 apples. As a survey of the kitchen is taken, we find 2 pie crusts and 10 apples. How many pies can be made? It is not difficult to see that we have enough ingredients to make 2 pies; that will consume the 2 crusts that we have, but will only use 8 of the 10 apples. In our chemical language, the recipe is our balanced chemical equation, the pie crust is the limiting reactant, and the apples represent the reactant in excess.

Now, let's apply what we have just learned in the kitchen to the following reaction:

$$N_2(g) + 3H_2(g) \rightarrow 2NH_3(g) \qquad\qquad \textbf{Equation 2}$$

If 2 mol of $N_2$ and 5 mol of $H_2$ are placed into a reaction vessel, how much $NH_3$ will be produced? Since moles are given, we can directly identify which reactant is the limiting reactant (since the limiting reactant controls how much product is generated) by using the molar relationships from the balanced chemical equation. The calculations used to determine the limiting reactant are shown below:

$$\left( 2 \text{ mol } N_2 \right) \times \left( \frac{3 \text{ mol } H_2}{1 \text{ mol } N_2} \right) = 6 \text{ mol } H_2 \text{ needed}$$

don't have enough $H_2$, so $H_2$ is LIMITING REACTANT

OR

$$\left( 5 \text{ mol } H_2 \right) \times \left( \frac{1 \text{ mol } N_2}{3 \text{ mol } H_2} \right) = 1.67 \text{ mol } N_2 \text{ needed}$$

have more than enough $N_2$, so $N_2$ is in EXCESS, and $H_2$ is LIMITING REACTANT

For each calculation, a conversion factor, determined from the coefficients in the balanced chemical equation, is used to reflect the numerical relationship between the moles of each reactant required for the reaction. Either calculation reveals the same fact: there is an excess of $N_2$, so $H_2$ is the limiting reactant. (*Note:* You only need to do one of the calculations shown—they both will allow you to determine unequivocally which reactant is the limiting reactant.)

Alternatively, one can determine the limiting reactant by comparing molar ratios based on the coefficients from the balanced equation. As shown in the following, on the left is the molar ratio defined by the balanced chemical equation, while the right side fixes one of the reactants at the given amount and allows us to determine how much of the other would be needed. So for the top calculation, the thought process can be expressed as such: "every 1 mol of $N_2$ will react with exactly 3 mol of $H_2$; if we want to react 2 mol of $N_2$ (given), then how much $H_2$ (X) would be needed?" Each equation is solved using cross multiplication.

$$\left(\frac{1 \text{ mol } N_2}{3 \text{ mol } H_2}\right) = \left(\frac{2 \text{ mol } N_2}{X \text{ mol } H_2}\right) \longrightarrow X = 6 \text{ mol } H_2 \text{ needed}$$

don't have enough $H_2$, so $H_2$ is LIMITING REACTANT

OR

$$\left(\frac{1 \text{ mol } N_2}{3 \text{ mol } H_2}\right) = \left(\frac{X \text{ mol } N_2}{5 \text{ mol } H_2}\right) \longrightarrow X = 1.67 \text{ mol } N_2 \text{ needed}$$

have more than enough $N_2$, so $N_2$ is in EXCESS, and $H_2$ is LIMITING REACTANT

Now with $H_2$ identified as the limiting reactant, we can use the conversion factor that relates the amount of $NH_3$ produced to the amount of $H_2$ consumed *or* the ratio method as follows:

$$\left(5 \text{ mol } H_2\right) \times \left(\frac{2 \text{ mol } NH_3}{3 \text{ mol } H_2}\right) = 3.33 \text{ mol } NH_3 \text{ produced}$$

OR

$$\left(\frac{3 \text{ mol } H_2}{2 \text{ mol } NH_3}\right) = \left(\frac{5 \text{ mol } H_2}{X \text{ mol } NH_3}\right) \longrightarrow X = 3.33 \text{ mol } NH_3 \text{ produced}$$

The answer from this calculation is called the *theoretical yield*, or the maximum amount of $NH_3$ that can be produced based on the starting materials and the reaction stoichiometry. The *actual yield*, the amount of product actually produced, is nearly always less than the theoretical yield for a number of possible reasons: the product is not totally recovered during the purification process; the reaction doesn't proceed completely due to reaction conditions; there might be evaporation of a volatile product such as a gas; or there may be a competing reaction (with its own chemical equation and stoichiometric relationships) where by-products are produced, thereby leaving less reactants available for the desired reaction. The percent yield (% yield) is a comparison of the actual to the theoretical yields and provides a measure of the efficiency of the reaction:

$$\% \text{ yield} = \frac{\text{actual yield}}{\text{theoretical yield}} \times 100 \qquad \qquad \textbf{Equation 3}$$

## THIS LAB

In this lab, you will predict and observe a limiting reactant during the reaction between copper(II) chloride dihydrate $[CuCl_2 \cdot 2H_2O]$ and metallic aluminum $[Al(s)]$:

$$3CuCl_2 \cdot 2H_2O(aq) + 2Al(s) \rightarrow 3Cu(s) + 2AlCl_3(aq) + 6H_2O$$

This is known as an oxidation-reduction (or "redox") reaction, where electrons are exchanged between the two metals as they change their oxidation states. $CuCl_2$ turns a light blue in aqueous solution due to the $Cu^{2+}(aq)$ ion, while aqueous $AlCl_3$ is colorless. By varying the $CuCl_2 \cdot 2H_2O$ and $Al(s)$ quantities and observing the resulting reactions, you will be able to visualize limiting and excess reactants. You will also use stoichiometric calculations to determine the theoretical yield of Cu(s) from one of the reactions, and then compare the actual yield to determine the percent yield.

## Procedure

1. Label two 250-mL beakers "1" and "2," respectively. Record the mass of beaker 1.
2. Measure out the following amounts of $CuCl_2 \cdot 2H_2O$, add to the beakers, and record the actual masses:
   - 0.50 g $CuCl_2 \cdot 2H_2O$ into beaker 1
   - 0.70 g $CuCl_2 \cdot 2H_2O$ into beaker 2
3. Add 50 mL deionized $H_2O$ to each of the beakers; gently mix the solutions until the copper salt is completely dissolved.
4. Measure out the following amounts of aluminum foil and add to the beakers in small pieces; record the actual masses:
   - 0.25 g Al(s) into beaker 1
   - 0.05 g Al(s) into beaker 2
5. Inspect the contents of each beaker and record all observations (e.g., colors, smell, bubbling, heat formation, etc.).
6. Stir the contents of each beaker periodically with a glass stirring rod and record any changes you observe.
7. Once the reactions are complete (how do you know this?), record the colors of the beaker contents and any other observations.
8. In beaker 1, if excess aluminum foil is still observed, then in a hood, add 6 M HCl in small portions until the foil is completely reacted.
9. After allowing the solid copper product to settle, decant the solution, being careful to not lose any of the copper.
10. Wash the copper solid with 15 mL of deionized water, let solid settle, and decant; repeat once.
11. Wash the copper solid with 10 mL of methanol, let solid settle, and decant.
12. In the hood, heat beaker 1 containing the copper solid on a hot plate at a low setting until dry. *Note:* Avoid heating at high temperatures for longer periods of time, which may cause the unwanted oxidation of the copper product.
13. After cooling, record the mass of beaker 1 and its contents.
14. Now, take a third 250-mL beaker and add 0.70 g $CuCl_2 \cdot 2H_2O$ (record actual mass) and 50 mL deionized $H_2O$.
15. Determine how much Al(s) is needed (i.e., the stoichiometric amount) in order to completely react all of the $CuCl_2$. Measure this amount out (record mass) and add it to beaker in small pieces.
16. Record your observations initially, during the reaction, and at the conclusion of the reaction.
17. Dispose of the contents of the beakers as indicated by your instructor.

## Data Collection

1. Mass of beaker 1: _____

2. Mass of $CuCl_2 \cdot 2H_2O$:      Beaker 1: _____

                                      Beaker 2: _____

3. Mass of Al(s):               Beaker 1: _____

                                      Beaker 2: _____

**4.** Observations for reactions:

|  | Beaker 1 | Beaker 2 |
|---|---|---|
| **Initial** (immediately after being mixed) | | |
| **Changes** during reaction | | |
| **After reaction** is complete | | |

**5.** Mass of beaker 1 + contents: _____

**6.** Mass of $CuCl_2 \cdot 2H_2O$:        Beaker 3: _____

**7.** Mass of Al($s$):        Beaker 3: _____

**8.** Observations for reaction:

|  | Beaker 3 |
|---|---|
| **Initial** (immediately after being mixed) | |
| **Changes** during reaction | |
| **After reaction** is complete | |

## Analysis

**9.** Beaker 1 calculations:

Moles of $CuCl_2 \cdot 2H_2O$: _____

Moles of Al($s$): _____

Moles of Cu($s$) produced (theoretical): _____

Mass of Cu($s$) produced (theoretical): _____

Mass of Cu($s$) produced (actual) = (#5) − (#1): _____

Percent yield of Cu($s$): _____

**10.** Beaker 2 calculations:

Moles of $CuCl_2 \cdot 2H_2O$: _____

Moles of $Al(s)$: _____

Moles of $Cu(s)$ produced (theoretical): _____

Moles of $AlCl_3$ produced (theoretical): _____

**11.** Beaker 3 calculations:

Moles of $CuCl_2 \cdot 2H_2O$: _____

Moles of $Al(s)$: _____

Moles of $Cu(s)$ produced (theoretical): _____

Moles of $AlCl_3$ produced (theoretical): _____

## Reflection Questions

1.  For the reactions in beakers 1 and 2, each had a total mass of reactants of ~0.75 g.
    (a) Which combination of reactants (beaker 1 or 2) will produce the largest total number of moles of products? Explain.

    (b) Which combination of reactants (beaker 1 or 2) will produce the largest total number of grams of products? Explain.

2.  For each reaction, what observation(s) indicated that the reaction was complete? Also, why did each reaction stop—be specific?
    (a) Beaker 1:

    (b) Beaker 2:

    (c) Beaker 3:

3. For each reaction, based on your calculations, which reactant was the limiting reactant? How did your observations for each reaction reinforce your answer?
   (a) Beaker 1:

   (b) Beaker 2:

   (c) Beaker 3:

4. Is it true that there is a limiting reactant present in *any* reaction that is run? Explain.

5. For the reaction in beaker 2, what experimental errors could have possibly contributed to your percent yield?

6. If all water was removed from beakers 1 through 3, list the chemical species that would be left in the beakers:
   (a) Beaker 1:

   (b) Beaker 2:

   (c) Beaker 3:

## Connection

Based on your experience in this lab, draw a connection to something in your everyday life or the world around you (something not mentioned in the background section):

# 9

# Qualitative Analysis: Testing the Solubility Rules

## Objectives

- Understand the solubility rules.
- Be able to write molecular, ionic, and net ionic equations.
- Identify cations and anions based on their solubility profile.
- Learn the proper techniques for the handling and disposal of chemicals.

## Materials Needed

- 13 mm $\times$ 100 mm test tubes (approximately 35)
- Beral pipettes
- Test tube rack
- Seven aqueous solutions (0.1 $M$) containing $K^+$, $Cu^{2+}$, $Ag^+$, $Ba^{2+}$, $Fe^{3+}$, $Mg^{2+}$, and $Pb^{2+}$ as nitrate salts (labeled A through G in liquid dropper bottles)
- Sodium bromide (NaBr), solid
- Sodium sulfate ($Na_2SO_4$), 0.1 $M$ aqueous solution
- Sodium phosphate ($Na_3PO_4$), 0.1 $M$ aqueous solution
- Sodium hydroxide (NaOH), 0.1 $M$ aqueous solution
- Unknown anion aqueous solution containing any one of the following in a liquid dropper bottle: $Br^-$, $Br^-/SO_4^{2-}$, $Br^-/PO_4^{3-}$, $Br^-/OH^-$, $SO_4^{2-}$
- Deionized water
- 10-mL graduated cylinder
- 25-mL Erlenmeyer flask
- Permanent marking pen

## Safety Precautions

- Gloves and safety goggles/glasses should be worn during this lab.
- Some of the metal cations used in this lab, in particular $Pb^{2+}$ and $Ba^{2+}$, are toxic by ingestion. Skin contact should be avoided.
- Dispose of waste solutions in designated waste containers as indicated by your instructor.

## QUALITATIVE ANALYSIS

Analysis is a particularly important part of chemistry. *Quantitative* analysis involves the determination of specific amounts or percentages of matter present in a sample, typically answering the question "how much?" Examples of quantitative analysis include gravimetric (lab 7) and volumetric (lab 10) analysis. *Qualitative* analysis, however, is not concerned with the amount, but simply the identity of matter present in a sample, answering the question "what's there?" Forensic science, with applications in arson and explosive detection, illegal drug analysis, toxicology, fingerprinting, crime scene investigation, etc., is often focused on qualitative analytical methods. In qualitative analysis, the identities of substances can be determined through observation of physical or chemical properties. In the former case, a characteristic color, fluorescence, melting point, boiling point, smell, etc. may be helpful in identification. Chemical properties refer to a chemical change that is specific to a substance. In other words, observing a chemical reaction that is characteristic of a particular substance identifies chemical properties. In this lab, you will be using both physical (primarily color and appearance) and chemical properties (precipitation reactions) to identify unknown cations and anions.

## THE NATURE OF IONIC COMPOUNDS

As a graduate student in 1884, Svante Arrhenius theorized that ionic compounds dissociate upon dissolving in water, resulting in solutions that conduct electricity. This simple idea, obvious today but revolutionary at the time, was not well received by his professors. In fact, Arrhenius barely passed the defense of his graduate dissertation, ironic in that this work eventually led to his receipt of the Nobel Prize in 1903!

What happens when an ionic compound or any strong electrolyte is dissolved in water? The answer is important in understanding the properties of solutions containing ions and in the chemical reactions that take place involving ionic species. Consider the formation of a solution by dissolving the water-soluble ionic compound potassium sulfate in water. As shown below, each formula unit of potassium sulfate dissociates to give two potassium cations and one sulfate ion separated in solution. These ions are stabilized through ion-dipole interactions with the polar water molecule.

$$K_2SO_4(s) \xrightarrow{H_2O} 2K^+(aq) + SO_4^{2-}(aq)$$

Not all ionic compounds are soluble in water. The solubility rules (Table 1) provide a summary of solubility patterns based on the types of ions that are present in a particular compound. However, the rules in Table 1 deserve minor clarification. While easy to use (e.g., AgCl is insoluble in water, but $AgNO_3$ is soluble), they present ionic compounds as either soluble or insoluble in water with no interpretation allowed for the degree of solubility. For example, silver sulfate, $Ag_2SO_4$, is listed as an exception to the rule that sulfate salts are generally soluble in water. Silver sulfate is certainly not *very* soluble in water, but approximately 5 g *will* dissolve in 1 L of water. Thus, it is perhaps best characterized as moderately soluble in water (a description that may prove useful to you in today's lab). In short, solubility is best defined in a quantitative way, as in the grams or moles of solute dissolved per liter of solution. Nevertheless, the solubility patterns introduced by the solubility rules are clear and important for you to know because they are valid for a wide range of cations and anions and allow for the rapid prediction of precipitation reactions.

## Table 1　The Solubility Rules for Ionic Compounds.

### Soluble Compounds

| Water-soluble Compounds | Insoluble Exceptions |
|---|---|
| Compounds containing an alkali metal cation ($Li^+$, $Na^+$, $K^+$, $Rb^+$, $Cs^+$) or ammonium ion ($NH_4^+$) | |
| Compounds containing the nitrate ion ($NO_3^-$), acetate ion ($C_2H_3O_2^-$) or chlorate ion ($ClO_3^-$) | |
| Compounds containing the chloride ion ($Cl^-$), bromide ion ($Br^-$) or iodide ion ($I^-$) | Compounds containing $Ag^+$, $Hg_2^{2+}$ or $Pb^{2+}$ |
| Compounds containing the sulfate ion ($SO_4^{2-}$) | *Compounds containing $Ag^+$, $Hg_2^{2+}$, $Pb^{2+}$, $Ca^{2+}$, $Sr^{2+}$, or $Ba^{2+}$ |

### Insoluble Compounds

| Water-insoluble Compounds | Soluble Exceptions |
|---|---|
| Compounds containing the carbonate ion ($CO_3^{2-}$), phosphate ion ($PO_4^{3-}$), chromate ion ($CrO_4^{2-}$), or sulfide ion ($S^{2-}$) | Compounds containing $Li^+$, $Na^+$, $K^+$, $Rb^+$, $Cs^+$, or $NH_4^+$ |
| Compounds containing the hydroxide ion ($OH^-$) | Compounds containing $Li^+$, $Na^+$, $K^+$, $Rb^+$, $Cs^+$, or $Ba^{2+}$ |

*While $Ag^+$ is typically listed as an insoluble exception for the sulfate ion, $Ag_2SO_4$ is in fact moderately soluble in water.

## PRECIPITATION REACTIONS

The mixing of two solutions to produce an insoluble product is characteristic of a precipitation reaction. As shown in Figure 1, a solution of sodium iodide is mixed with lead(II) nitrate to give the bright yellow precipitate lead(II) iodide. From the solubility rules, this precipitation reaction is readily predictable, as the reactants are listed as water soluble and the product lead(II) iodide is not. This precipitation reaction can be described by a balanced *molecular* equation in which the neutral formulas of the reactants and products are included. However, the molecular form of this equation does not describe the soluble ionic compounds in the form in which they are actually present, namely, free ions in solution. As such, an alternate equation, called an *ionic* equation, reflects the true nature of each chemical species by showing all strong electrolytes (e.g., soluble ionic compounds) as dissociated ions. Note that the lead(II) iodide is not shown dissociated in the ionic equation because it is the precipitate and is present as the intact ionic compound in the reaction mixture (see the yellow precipitate in Figure 1). Inspection of the ionic equation reveals that the $Na^+$ and $NO_3^-$ ions are unchanged during the reaction. They are deemed *spectator ions* and are omitted in the *net ionic equation*, which describes only the chemical species that have participated in the reaction.

*Molecular*　　　　　　　　　　$Pb(NO_3)_2(aq) + 2NaI(aq) \rightarrow PbI_2(s) + 2NaNO_3(aq)$

*Ionic*　　　$Pb^{2+}(aq) + 2NO_3^-(aq) + 2Na^+(aq) + 2I^-(aq) \rightarrow PbI_2(s) + 2Na^+(aq) + 2NO_3^-(aq)$

*Net ionic equation*　　　　　　　$Pb^{2+}(aq) + 2I^-(aq) \rightarrow PbI_2(s)$

**Figure 1** **The formation of $PbI_2(s)$ from the reaction of $NaI(aq)$ with $Pb(NO_3)_2(aq)$.**

## THIS LAB

The first section of this lab requires the preparation of a 0.2 $M$ NaBr solution to be used for cation analysis. The symbol "$M$" refers to *molarity*, a quantitative measure of solution composition. Molarity is defined as the moles of solute dissolved in 1 L of solution, or $M$ = moles of solute per liter of solution. The more concentrated the solution, the higher the molarity. In addition to describing accurately the composition of a solution, molarity has great value as a conversion factor, allowing for the determination of moles present in a given volume of solution or the determination of volume that contains a specific molar amount of solute. Your solution need only be *approximately* 0.2 $M$ NaBr, as the qualitative analysis in this lab simply requires the presence of bromide in solution.

In the second section of the lab, you will be given seven solutions, labeled A through G, each containing one of the following cations: $K^+$, $Cu^{2+}$, $Ag^+$, $Ba^{2+}$, $Fe^{3+}$, $Mg^{2+}$, and $Pb^{2+}$. Your task is to use the prepared NaBr solution and provided solutions of $Na_2SO_4$, $Na_3PO_4$, and NaOH, along with an understanding of the solubility rules, to identify the cations present in A through G through a series of precipitation reactions.

In the final section, you will use your conclusions concerning the identity of A–G to solve the following problem. For a century (well into the middle of the 1900s), the anion bromide was valued as a sedative and a treatment for epilepsy. Due to medical advances and side effects associated with its use, bromide has little medical use at the present time. However, it has found continued use in veterinary medicine for treating epileptic dogs. Interestingly, because of its former application as a sedative, "bromide" has infiltrated the English language to describe a story or person that is dull or boring. As a student researcher in a laboratory interested in understanding the action of bromide in living systems, you have found that one of your solutions has particularly interesting activity. Unfortunately, the contents of this solution may contain any of the five following possibilities: $Br^-$, $Br^-$/ $SO_4^{2-}$, $Br^-$/$PO_4^{3-}$, $Br^-$/$OH^-$, or $SO_4^{2-}$. You would like to attribute the interesting biological activity to the presence of bromide alone, but other anions that might be present could also be responsible for the activity. As such, you must analyze your solution to determine all of the anions that are present. Does your solution contain only bromide?

### References

1. (1966). *Nobel Lectures, Chemistry 1901–1921*, Elsevier Publishing Company, Amsterdam.
2. Rodgers GE. (2012). *Descriptive Inorganic, Coordination, and Solid-State Chemistry.* 3d ed., Brooks/Cole, Cengage Learning, Belmont, CA.

## Procedure

### PREPARATION OF A 0.2 *M* NaBr SOLUTION

1. Determine and record the amount of solid NaBr needed to prepare 10 mL of a 0.2 *M* solution.
2. Measure and record the mass of a sample of NaBr that is approximately equal to that determined in step 1. (*Note: You need not measure a mass identical to that determined from step 1. The target solution concentration of 0.2 M contains only one significant figure!*)
3. Add the solid NaBr to a 25-mL Erlenmeyer flask and add 10 mL of water to make an approximately 0.2 *M* solution. (*Note: If the measured mass of NaBr in step 2 differs significantly from that required in step 1, then you need to adjust the volume of water accordingly to make a solution concentration of approximately 0.2 M.*)
4. Label one test tube "0.2 *M* NaBr" using the permanent marking pen and then fill it ¾ full with the prepared NaBr solution.
5. Label three test tubes with "0.1 *M* $Na_2SO_4$", "0.1 *M* $Na_3PO_4$" and "0.1 *M* NaOH", respectively, and fill them each ¾ full from the appropriate provided solutions.

### CATION ANALYSIS

6. Arrange seven sets of four test tubes each in a test tube rack to match the grid shown in the data collection section of this lab. Each set of four tubes should be labeled A–G.
7. Add 10 to 12 drops of the aqueous solution containing cation "A" to the four "A" labeled test tubes. Proceed in a similar way to fill the remaining sets of four test tubes with solutions "B", "C", "D", "E", "F" and "G", respectively.
8. Progressing left to right across the grid, add 10 to 12 drops of NaBr(*aq*) using a beral pipette to each of samples A through G. Record all observations (no change, precipitation, precipitate color, solution color, etc.) on the grid provided in the data collection section. (*Note: Do not use the same beral pipette for different solutions.*)
9. Proceed as in step 8 by adding $Na_2SO_4$(*aq*) to one sample each of A–G and record observations on the grid.
10. Proceed as in step 8 by adding $Na_3PO_4$(*aq*) to one sample each of A–G and record observations on the grid.
11. Proceed as in step 8 by adding NaOH(*aq*) to one sample each of A–G and record observations on the grid.
12. Based on the solubility rules and your results, identify cations A through G as $K^+$, $Cu^{2+}$, $Ag^+$, $Ba^{2+}$, $Fe^{3+}$, $Mg^{2+}$, or $Pb^{2+}$. Each cation solution corresponds uniquely to one of these cations.
13. Write molecular, ionic, and complete ionic equations for each of the reactions that produced a precipitate.

### ANION ANALYSIS

14. Obtain an unknown solution containing one of the following five possibilities: $Br^-$, $Br^-/SO_4^{2-}$, $Br^-/PO_4^{3-}$, $Br^-/OH^-$ or $SO_4^{2-}$. Add 10 to 12 drops of your unknown solution to a test tube.
15. Add 10 to 12 drops of one of the cation solutions that you deem necessary to help identify your unknown to the test tube containing the unknown. Record all observations and conclusions.
16. Repeat steps 14 and 15 as necessary until you have identified your unknown. Identify the composition of the unknown.

## Data Collection

### PREPARATION OF A 0.2 *M* NaBr SOLUTION

1. Required mass of NaBr, step 1:  _____
2. Measured mass of NaBr, step 2:  _____

**CATION ANALYSIS**

3.

| | Cation A | Cation B | Cation C | Cation D | Cation E | Cation F | Cation G |
|---|---|---|---|---|---|---|---|
| $Br^-$ | | | | | | | |
| $OH^-$ | | | | | | | |
| $SO_4^{2-}$ | | | | | | | |
| $PO_4^{3-}$ | | | | | | | |

## ANION ANALYSIS

**4.** *Observations* and *conclusions* for each individual anion analysis (in your conclusions, clearly state which anions you eliminated or identified as present and why), step 15:

Test tube 1: _____ was added to unknown.

             (list cation)

Observations:

Conclusions:

Test tube 2: _____ was added to unknown.

             (list cation)

Observations:

Conclusions:

Test tube 3: _____ was added to unknown.

             (list cation)

Observations:

Conclusions:

Test tube 4: _____ was added to unknown.

             (list cation)

Observations:

Conclusions:

Test tube 5: _____ was added to unknown.

          (list cation)

Observations:

Conclusions:

Test tube 6: _____ was added to unknown.

          (list cation)

Observations:

Conclusions:

## *Analysis*

## CATION ANALYSIS

5. Identification of cation samples, step 12:

_____   _____   _____   _____   _____   _____   _____

  **A**       **B**       **C**       **D**       **E**       **F**       **G**

6. Molecular, ionic, and net ionic equations (based on the results from your grid), step 13:
   (*Note: Write equations only for reactions that produced a precipitate.*)

**Reaction 1**

Molecular:

Total Ionic:


Net Ionic:


## Reaction 2

Molecular:


Total Ionic:


Net Ionic:


## Reaction 3

Molecular:


Total Ionic:


Net Ionic:

**Reaction 4**

    Molecular:

    Total Ionic:

    Net Ionic:

**Reaction 5**

    Molecular:

    Total Ionic:

    Net Ionic:

**Reaction 6**

    Molecular:

    Total Ionic:

Net Ionic:

## Reaction 7

Molecular:

Total Ionic:

Net Ionic:

## Reaction 8

Molecular:

Total Ionic:

Net Ionic:

## Reaction 9

Molecular:

Total Ionic:

Net Ionic:

**Reaction 10**

Molecular:

Total Ionic:

Net Ionic:

**Reaction 11**

Molecular:

Total Ionic:

Net Ionic:

**Reaction 12**

    Molecular:

    Total Ionic:

    Net Ionic:

**Reaction 13**

    Molecular:

    Total Ionic:

    Net Ionic:

**Reaction 14**

    Molecular:

    Total Ionic:

Net Ionic:

**Reaction 15**

Molecular:

Total Ionic:

Net Ionic:

**Reaction 16**

Molecular:

Total Ionic:

Net Ionic:

## ANION ANALYSIS

**7.** Identification of anion(s) in unknown, step 16:

_____          _____

## Reflection Questions

1. In the first part of this lab, you prepared a NaBr solution that is approximately 0.2 $M$. Considering that molarity is defined as moles of NaBr per *liter of solution*, give possible errors that are present in the procedures that will cause your actual molarity to be different from that calculated. (*Note: Because this lab involved qualitative analysis, the actual molarity of your solution did not need to be known accurately.*)

2. The mixing of aqueous solutions of chromium(III) nitrate and sodium sulfide produces a brownish black powder as a precipitate.
   (a) Write a balanced equation that describes the dissociation of chromium(III) nitrate in water.

   (b) Write the balanced net ionic equation that describes the reaction.

   (c) What are the spectator ions?

3. For the following reaction, $MgBr_2(aq) + CuSO_4(aq)$,
   (a) Write the balanced molecular equation. (*Note: You must include phase labels.*)

   (b) What is the color of the precipitate?

   (c) What is the color of the final solution?

4. Consider a solution that contains a mixture of $Ba^{2+}$ and $Fe^{3+}$.
   (a) What anion would you add to only precipitate $Fe^{3+}$?

   (b) What anion would you add to only precipitate $Ba^{2+}$?

## Connection

Based on your experience in this lab, draw a connection to something in your everyday life or the world around you (something not mentioned in the background section):

# 10

# Titration I. Determination of an Unknown Diprotic Acid Through Volumetric Analysis

## Objectives

- Understand the techniques and equipment associated with titrations.
- Apply stoichiometric principles for molarity and molar mass determinations.
- Reinforce the importance of significant figures in measurement and calculations.

## Materials Needed

- 50-mL burette
- Burette stand
- Funnel
- 6 $M$ sodium hydroxide (NaOH) stock solution
- Potassium hydrogen phthalate (KHP), dried
- Unknown diprotic acid (succinic, malic, or tartaric acid)
- Deionized water*
- Phenolphthalein indicator solution
- 250-mL Erlenmeyer flask(s)
- 250-mL plastic container with lid
- 50-mL graduated cylinder
- Laboratory balance

## Safety Precautions

- Safety goggles/glasses should be worn during this lab.
- NaOH is a strong base that can cause burns. Avoid contact with body tissues.

## Background

### VOLUMETRIC ANALYSIS

Quantitative analytical methods for determining accurate concentrations of solutions or the moles of a substance present in a sample are extremely important in chemistry. The most common are based either on mass measurements (gravimetric analysis—see lab 7) or volume measurements (volumetric analysis or

---

*For best results, the deionized water should be boiled for 5 minutes to minimize the concentration of dissolved carbon dioxide. Basic solutions absorb and react with carbon dioxide from the air, which reduces the concentration of hydroxide in solution over time. Upon cooling, the boiled, deionized water should be kept in a sealed container and used throughout this lab whenever deionized water is needed.

*titrations*). Acid/base neutralization reactions can be conveniently monitored by titration. In an acid/base titration experiment, two solutions, each containing a separate reactant (acid in one, base in the other), are mixed in a controlled manner. The concentration of one of the solutions or molar amount of one of the reactants in solution is accurately known (the *standard* solution), whereas the other solution's concentration is unknown. Through careful addition of one reactant to the other, the reaction can be stopped precisely when equimolar amounts of acid and base have been added such that neither reactant is present in excess; this is known as the *equivalence point* of the titration. The proviso here is that the experimentalist must know when the equivalence point in the titration has occurred. This is accomplished through the addition of an acid/base *indicator*, a chemical substance whose color (and structure) changes at the equivalence point of the titration. By knowing the amount of standard solution needed to reach the equivalence point, one can determine the amount of the unknown substance present through stoichiometric calculation using the balanced chemical equation that describes the acid/base reaction.

A successful titration experiment requires an understanding of the equipment needed and the relevant chemistry. For accurate and precise results, a burette is used as the volume measuring device (see Figure 1). A standard 50-mL burette measures volume to the hundredths position, or two places beyond the decimal point. It is critical that this level of precision be reflected in each volume measurement; calculations that involve these measurements will therefore result in answers that reflect this number of significant figures. The use of the common Erlenmeyer flask as the receiving vessel is also noteworthy, since it is well-suited for "swirling." Periodically during the titration, the contents of the Erlenmeyer flask are mixed by gently swirling the solution. Further, because of its conical shape, the possibility of losing small amounts of solution due to splattering upon addition from the burette is minimized.

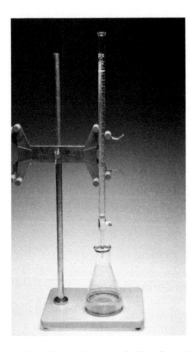

**Figure 1** The typical equipment needed for a titration, a burette (with a burette stand), and an Erlenmeyer flask, each containing a separate reactant. An indicator is initially added to the solution in the Erlenmeyer flask.

Indicators are weak acids that have distinct colors in an acidic versus basic environment. As noted previously, they are used in the titration experiment to signify that the endpoint of the titration has been reached. They do so by undergoing their own acid/base reaction, resulting in a color change, but only after the primary acid in the Erlenmeyer flask has been neutralized. As such, the goal of the experimentalist is to perform a titration until the faintest appearance of a persistent change in color (typically lasting more than 25 sec) is present. Too intense a color indicates that the endpoint of the titration has been exceeded with the consequent introduction of error in the result. For example, as shown in Figure 2 (left), addition of base from the burette to acid in the Erlenmeyer flask is accompanied by the appearance of a pink color. In this particular case, the indicator present in the acid solution is phenolphthalein. Phenolphthalein is colorless in an acidic to neutral environment, but pink in a more basic environment. With swirling, the pink color will dissipate until enough base has been added to neutralize the acid that is present. At this point, the addition of a small amount of base (perhaps half of one drop) gives rise to the image in Figure 2 (right)—a faint, persistent pink color reflecting the endpoint of the titration.

**Figure 2**   As base is added to the acid, the phenolphthalein indicator turns pink in color (left) until the endpoint of the titration is reached (right).

The final piece to understanding acid/base titrations is the need for a balanced chemical equation that describes the reaction between the acid and base. The coefficients in the balanced equation describe the exact relationship between the moles of acid and moles of base that have reacted at the equivalence point. For example, as shown below, the reaction involving NaOH and HCl clearly shows that 1 mole of NaOH is needed to react exactly with 1 mole of HCl; or, if a solution containing a known number of moles of HCl is titrated with a NaOH solution, then, at the endpoint of the titration, the moles of added NaOH must be equal to the moles of HCl present. In contrast, it takes 2 moles of NaOH to completely neutralize or reach the final endpoint in the titration of the diprotic acid $H_2SO_4$.

$$HCl(aq) + NaOH(aq) \rightarrow H_2O(l) + NaCl(aq)$$
$$H_2SO_4(aq) + 2NaOH(aq) \rightarrow 2H_2O(l) + Na_2SO_4(aq)$$

## THIS LAB

### Preparing a NaOH Standard Solution

In the first part of this lab, a NaOH solution will be prepared with an approximate concentration of 0.25 $M$. This solution will then be standardized, or have its concentration determined accurately, by titrating against the monoprotic acid standard KHP, potassium hydrogen phthalate ($KHC_8H_4O_4$, molar mass = 204.22 g mol$^{-1}$).

KHP is a colorless solid, available in pure form, with good water solubility (though only slightly hygroscopic). Thus, a mass of dry KHP will be accurately measured on an analytical balance and then transferred to an Erlenmeyer flask. Addition of deionized water and a few drops of phenolphthalein indicator results in a solution that will be titrated by the prepared NaOH solution. The balanced chemical equation describing the reaction of KHP with NaOH is shown below with $K^+$ and $Na^+$ serving as spectator ions:

$$KHC_8H_4O_4(aq) + NaOH(aq) \rightarrow KNaC_8H_4O_4(aq) + H_2O(l)$$

At the equivalence point of the titration, the moles of added NaOH will equal the moles of KHP present initially since the stoichiometric relationship between the two reactants is 1:1 (the moles of KHP can be readily determined from the measured mass and its molar mass). Then, dividing the moles of NaOH needed to neutralize the KHP by the volume of base delivered from the burette will give the actual concentration of the NaOH solution. Once the concentration of the NaOH solution is accurately known, it will be used as a standard solution in a titration to identify an unknown acid in the second part of this lab.

## Identification of an Unknown Diprotic Acid

You will be assigned one of three possible unknowns: succinic acid, malic acid, or tartaric acid. Their structures and molar masses are given in Figure 3. All are related by having a four-carbon backbone that terminates at each end in acidic carboxyl groups (shown in red). Each is a member of a class of organic compounds known as carboxylic acids. Carboxylic acids and their derivatives are ubiquitous in chemistry and biology. For example, acetic acid is a carboxylic acid and the carboxyl group present in all amino acids inspires their name.

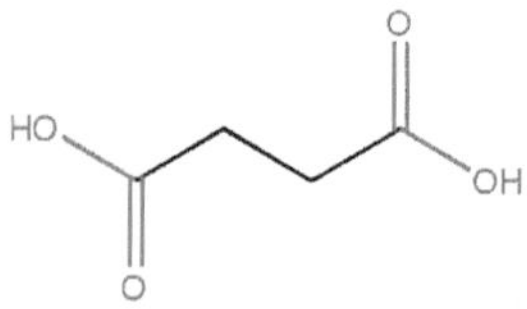

succinic acid
$C_4H_6O_4$
118.09 g mol$^{-1}$

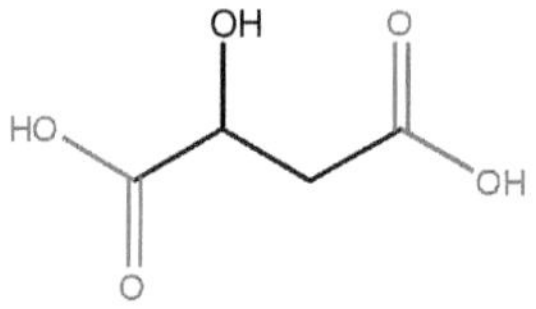

malic acid
$C_4H_6O_5$
134.09 g mol$^{-1}$

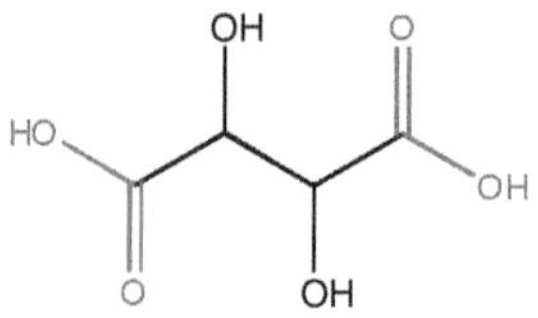

tartaric acid
$C_4H_6O_6$
150.09 g mol$^{-1}$

**Figure 3**  **Possible unknown diprotic acids used in this lab.**

All three of the potential unknowns are naturally occurring and biologically relevant acids. Succinic acid and malic acid are part of the Krebs cycle, a series of chemical reactions that all cells use to produce energy from carbohydrates, fats, and proteins. Tartaric acid is most notably present in grapes and bananas. Malic acid is present in most fruits, particularly apples, and is responsible for their tartness. Both tartaric acid and malic acid are the primary acid contributors to wine. Like acids in general, each of these substances is characterized by a sour taste. In fact, malic and tartaric acid are used in the food industry as souring agents (acidulants), with malic acid being the common source of "sour" in sour to extremely sour candies (see Figure 4).

To identify your unknown, you will need to identify a characteristic property that distinguishes your sample. All three acids are colorless and crystalline. Because they each contain two carboxyl groups, they are all diprotic. And as indicated, they are all sour (*Caution: never taste any chemicals used in a chemistry lab setting*). However, a distinguishing feature that links to their unique structures is their respective molar masses. By titrating known masses of your assigned unknown with your standardized NaOH solution, the moles of unknown in your sample can be determined. Dividing the measured mass by the moles as determined from titration will give the molar mass of your unknown. *But, be careful, each unknown is diprotic.*

**Figure 4  Sour candies typically contain varying amounts of malic acid.**

## Procedure

### PREPARATION OF A 0.25 *M* NaOH SOLUTION

1. Determine the volume of 6 *M* NaOH stock solution needed to prepare 250 mL of a 0.25 *M* NaOH solution. Show your calculations and record your result.
2. Transfer the volume of 6 *M* NaOH stock solution determined in step 1 to a 250-mL plastic bottle and dilute using deionized water to a final volume of approximately 250 mL. Swirl the solution, cap, and label the bottle. *Note: Basic solutions absorb and react with carbon dioxide from the air, which reduces the concentration of hydroxide in solution over time. Keep the NaOH solution in a closed container.*

### STANDARDIZATION OF THE 0.25 *M* NAOH SOLUTION

3. Measure 1.0 to 1.2 g of dried potassium hydrogen phthalate (KHP) on weighing paper and transfer it to a labeled 250-mL Erlenmeyer flask. Record the mass of KHP.
4. Dissolve the KHP using 50 mL of deionized water.
5. Add two to three drops of the phenolphthalein solution.
6. Using a funnel, rinse a clean 50-mL burette with three 5-mL portions of your standardized NaOH solution, making sure to pass the solution through the burette tip. Dispose of the waste NaOH solution in the appropriately labeled waste container.
7. Fill the burette with the NaOH solution using a funnel. The solution level should be near the top of the scale. Check that no air bubbles are trapped in the burette tip. Remove the funnel and record the initial volume of the solution to the correct precision that the burette allows.
8. Calculate and record the volume of base solution expected to reach the endpoint of your titration. Assume a 0.25 *M* concentration (NaOH) for this calculation.

9. Titrate the KHP by adding the NaOH solution with swirling of the Erlenmeyer flask to ensure complete and timely mixing of the two solutions. The tip of the burette should be below the rim of the Erlenmeyer flask to prevent the loss of titrant. The NaOH solution may be initially added fairly rapidly based on your calculation in step 8. When the volume of added base is a few milliliters less than calculated to reach the endpoint, decrease the addition rate of the NaOH. As the endpoint of the titration nears, the pink color of the indicator will become more persistent. At this point in the titration, the base should be added a single drop at a time. The sides of the flask and the burette tip (if a drop or half drop of base is attached) may be rinsed with deionized water. The endpoint of the titration is achieved when the indicator remains a faint pink color for at least 25 sec. If unsure of the endpoint being reached, then record the volume of NaOH delivered and add another ½ to 1 drop of NaOH solution by, again, rinsing the drop off of the burette tip with deionized water. If this results in too intense a color, then the endpoint has been passed and the volume previously recorded should be used for calculations. Once the endpoint has been reached, record the final volume of base in the burette.

   *Important: good titration technique will result consistently in endpoints that are the faintest, but persistent pink in color and ensure the best data.*

10. Refill the burette and repeat the titration with two additional KHP samples (i.e., repeat steps 3 through 9, with the omission of step 6).

11. Calculate the molarity of the NaOH solution for each of your trials. The results for the three trials should be within 0.0025 *M* of each other. If necessary, you may need to perform additional titrations.

12. Average the three molarity measurements that are closest or most precise and report that value as the actual concentration of the NaOH solution.


## UNKNOWN IDENTIFICATION: DETERMINATION OF MOLAR MASS

13. Measure 0.4 g of the unknown diprotic acid on weighing paper and transfer it to a labeled 250-mL Erlenmeyer flask. Record the mass of unknown used.

14. Dissolve the unknown acid using 50 mL of deionized water.

15. Add two to three drops of the phenolphthalein solution.

16. Fill the burette with the NaOH solution using a funnel, record the initial volume of base, and titrate the unknown solution as described in step 9. Record the final volume of base in the burette.

17. Repeat steps 13 through 16 with two additional samples of the unknown diprotic acid.


## Data Collection


## PREPARATION OF A 0.25 *M* NaOH SOLUTION

1. Volume of 6 *M* NaOH solution needed (show calculations):   _____

## STANDARDIZATION OF THE 0.25 *M* NaOH SOLUTION

**Trial 1**

2. Mass of KHP: _____

3. Initial volume of NaOH solution, $V_i$: _____

4. Volume of base solution expected to reach the endpoint
   (step 8, show calculations):



   _____

5. Final volume of NaOH solution, $V_f$: _____

**Trial 2**

6. Mass of KHP: _____

7. Initial volume of NaOH solution: _____

8. Final volume of NaOH solution: _____

**Trial 3**

9. Mass of KHP: _____

10. Initial volume of NaOH solution: _____

11. Final volume of NaOH solution: _____

**Trial 4 (if necessary)**

12. Mass of KHP: _____

13. Initial volume of NaOH solution: _____

14. Final volume of NaOH solution: _____

**Trial 5 (if necessary)**

15. Mass of KHP: _____

16. Initial volume of NaOH solution: _____

17. Final volume of NaOH solution: _____

## UNKNOWN IDENTIFICATION: DETERMINATION OF MOLAR MASS

**Trial 1**

18. Mass of unknown: _____

19. Initial volume of NaOH solution, $V_i$: _____

20. Final volume of NaOH solution, $V_f$: _____

**Trial 2**

21. Mass of unknown: _____

22. Initial volume of NaOH solution: _____

23. Final volume of NaOH solution: _____

**Trial 3**

24. Mass of unknown: _____

25. Initial volume of NaOH solution: _____

26. Final volume of NaOH solution: _____

## Analysis

### STANDARDIZATION OF THE 0.25 *M* NaOH SOLUTION

**Trial 1**

27. Moles of KHP (show calculations): _____

28. Moles of NaOH added at endpoint: _____

29. Volume of NaOH solution dispensed ($V_f - V_i$): _____

30. Molarity of NaOH solution (show calculations): _____

**Trial 2**

31. Moles of KHP: _____

32. Moles of NaOH added at endpoint: _____

33. Volume of NaOH solution dispensed ($V_f - V_i$): _____

34. Molarity of NaOH solution: _____

**Trial 3**

35. Moles of KHP: _____

36. Moles of NaOH added at endpoint: _____

37. Volume of NaOH solution dispensed ($V_f - V_i$): _____

38. Molarity of NaOH solution: _____

### Trial 4 (if necessary)

**39.** Moles of KHP: _____

**40.** Moles of NaOH added at endpoint: _____

**41.** Volume of NaOH solution dispensed ($V_f - V_i$): _____

**42.** Molarity of NaOH solution: _____

### Trial 5 (if necessary)

**43.** Moles of KHP: _____

**44.** Moles of NaOH added at endpoint: _____

**45.** Volume of NaOH solution dispensed ($V_f - V_i$): _____

**46.** Molarity of NaOH solution: _____

**47.** Actual molarity of the NaOH solution

(average of the three most precise trials): _____

**48.** Range of the three most precise trials: _____

## UNKNOWN IDENTIFICATION: DETERMINATION OF MOLAR MASS

**49.** Balanced equation describing the titration of the unknown diprotic acid ($H_2A$) with NaOH:

### Trial 1

**50.** Volume of NaOH solution dispensed ($V_f - V_i$): _____

**51.** Moles of NaOH added at endpoint = (#50) × (#47): _____

**52.** Moles of unknown diprotic acid titrated (show calculations): _____

**53.** Molar mass of unknown acid = $\dfrac{(\#18)}{(\#52)}$ : _____

### Trial 2

**54.** Volume of NaOH solution dispensed ($V_f - V_i$): _____

**55.** Moles of NaOH added at endpoint = (#54) × (#47): _____

**56.** Moles of unknown diprotic acid titrated: _____

**57.** Molar mass of unknown acid = $\dfrac{(\#21)}{(\#56)}$ : _____

**Trial 3**

58. Volume of NaOH solution dispensed ($V_f - V_i$): _____

59. Moles of NaOH added at endpoint = (#58) × (#47): _____

60. Moles of unknown diprotic acid titrated: _____

61. Molar mass of unknown acid $= \dfrac{(\#24)}{(\#60)}$: _____

62. Average molar mass of unknown acid (average of the three trials) $= \dfrac{[(\#53) + (\#57) + (\#61)]}{3}$:

_____

63. Range of the three trials: _____

64. Identification of unknown acid: _____

65. % error $= \dfrac{|(\text{known molar mass of acid}) - (\text{calculated molar mass of acid})|}{(\text{known molar mass of acid})} \times 100$

$= \dfrac{|(\text{known molar mass of acid}) - (\#62)|}{(\text{known molar mass of acid})} \times 100$

_____

## Reflection Questions

1. Citric acid, $C_6H_8O_7$, is a triprotic acid found in a variety of fruits (e.g., lemons and limes). It, like acids in general and the unknowns in this lab, has a quite sour taste.
   (a) Write the balanced chemical equation describing the neutralization of citric acid with NaOH.

   (b) What volume of 0.3244 $M$ NaOH is needed to completely neutralize 0.266 g of citric acid?

2. Acetic acid, $CH_3COOH$ or $HC_2H_3O_2$, is a weak monoprotic acid that contains, like the unknowns in this lab, the carboxyl group. It is responsible for the sour taste of vinegars.
   (a) Draw the Lewis structure of acetic acid and identify the acidic proton.

   (b) Write the balanced chemical equation describing the neutralization of acetic acid with sodium hydroxide.

   (c) To determine the percent acetic acid (mass percent) in a vinegar sample, you perform a titration using a 0.2508 $M$ NaOH standard solution. A 20.62-mL volume of the NaOH solution is required to reach the endpoint in the titration of a 7.248-g sample of vinegar. What is the mass percent of acetic acid in this vinegar sample?

3. How would the following errors affect the molar mass determination (too high, too low, or no effect) in the second part of this lab? Explain your answers.
   (a) After the burette has been filled with the NaOH solution and the initial volume of base has been recorded, a drop of NaOH solution falls from the tip of the funnel into the burette as the funnel was being removed. The titration of the unknown diprotic acid then proceeds.

   (b) Your standardized NaOH solution is left standing in air for a week, because the cap was left off its plastic container, and then used for the titration of the unknown acid. Assume that there is minimal to no loss in the total volume of the solution due to evaporation over the week.

(c) You titrated your unknown sample to a perfect endpoint, as indicated by a faint pink color due to the phenolphthalein indicator. However, you notice a drop of the NaOH solution remains hanging off the tip of the burette.

## Connection

Based on your experience in this lab, draw a connection to something in your everyday life or the world around you (something not mentioned in the background section):

# 11

# The Activity Series: On the Chemistry of Metals

## Objectives

- Learn the activity series.
- Be able to describe and balance simple redox reactions.

## Materials Needed

- 13 mm $\times$ 100 mm test tubes (17)
- Beral pipettes
- Test tube rack
- 0.1 $M$ copper(II) sulfate ($CuSO_4$)
- 0.1 $M$ zinc(II) sulfate ($ZnSO_4$)
- 0.1 $M$ silver nitrate ($AgNO_3$)
- 6 $M$ hydrochloric acid (HCl)
- Cu wire
- Ag wire
- Fe metal*
- Zn (mossy, small pieces)
- Mg ribbon*
- 25-mL Erlenmeyer flasks (4)
- Wire snips or cutters
- Permanent marking pen

## Safety Precautions

- Gloves and safety goggles/glasses should be worn during this lab.
- 6 $M$ hydrochloric acid is a concentrated, strong acid; avoid contact with body tissues. Use it in a safety hood.
- Hydrogen gas is generated upon treatment of some metals with acid; as hydrogen is flammable, keep all heat and flames away from your reaction vessel.
- Dispose of waste solutions in designated waste containers as indicated by your instructor.

*Ideally, pretreated with 3 $M$ HCl for 5 (for Mg) to 15 (for Fe) sec, followed by thorough washing with deionized water, to ensure a pristine surface.

## Background

Over three-quarters of the known elements can be classified as metals. In metallic form, these elements are typically shiny and silver in color with good thermal and electrical conductivities. They can be drawn into wires (ductile) and pounded into sheets (malleable). While these properties are generally shared, there are variances with each specific metal that give rise to unique applications and properties. For example, silver and copper are among the best electrical conductors and are used accordingly. Gold is distinctly colored, is relatively soft, and is readily formed into ornamental objects and jewelry and, as Rutherford demonstrated in his gold foil experiment, very thin sheets. Tungsten's hardness and very high melting point suits it well for use in filaments in the high-temperature confines of lightbulbs. Situated to the left in the periodic table, the metals are the more easily oxidized (loss of electrons) of the elements, resulting in the formation of cations and ionic compounds. In Earth's oxygen rich atmosphere, an oxidizing environment, the majority of metallic elements are found in nature in cationic form as components in rocks, dirt, sand, minerals, etc. Further, the shiny, silvery appearance of some metals is masked by the ready formation of metal oxides on their surface. These oxides can be colored, as in the burnt orange rust formation on iron (see Figure 1), transparent as in aluminum oxide on aluminum, or responsible for a dull gray appearance as in sodium. Some metal surfaces, like those on chromium and aluminum, are protected by a thin, but dense oxide coating, whereas others, as on iron, are covered with a porous oxide, allowing for continual attack.

**Figure 1** **An iron planter with considerable oxidation as evidenced by the rusty appearance.**

Like other properties, the common ability to lose electrons is not equal among the metals. Some, like the alkali metals, lose electrons easily, while others, such as gold and platinum, do not. The ease with which a particular metal oxidizes (reacts through loss of electrons) is summarized in the activity series. Sodium metal is more "active" than copper and, therefore, is placed higher in the activity series.

$$\text{More active} \qquad Na \rightarrow Na^+ + e^-$$
$$\text{Less active} \qquad Cu \rightarrow Cu^{2+} + 2\,e^-$$

These two reactions are considered *half-reactions*, in that they each describe only an oxidation process, or the loss of electrons. In reality, these "lost" electrons must go somewhere. In fact, they are transferred to another reacting species in a corresponding reduction process. As such, oxidation and reduction events are coupled in that one substance gains electrons as another substance supplies them. Reactions that involve the transfer of electrons between reactants are called oxidation-reduction (or redox) reactions.

As an example, consider the reaction of cobalt metal with an aqueous solution of silver ions described by the overall balanced Equation 1. The two relevant half-reactions are denoted as either oxidation or reduction half-reactions:

$$2Ag^+(aq) + Co(s) \rightarrow 2Ag(s) + Co^{2+}(aq) \qquad\qquad \textbf{Equation 1}$$

*Half-reactions:*

Oxidation:$\quad\quad\quad\quad\quad\quad\quad\quad$ $Co(s) \rightarrow Co^{2+} + 2e^-$

Reduction:$\quad\quad\quad\quad\quad\quad\quad\quad$ $Ag^+ + e^- \rightarrow Ag(s)$

Note that oxidation half-reactions list electrons as products because the reactant is oxidized or loses electrons. The reactant in the oxidation half-reaction is the *reducing agent* in a redox reaction or the source of electrons that cause reduction. Reduction half-reactions necessarily include electrons as reactants. The reactant in the reduction half-reaction is therefore the *oxidizing agent*, or the chemical species that causes oxidation in a redox reaction. Each of these half-reactions is balanced as written, but neither takes place alone. They are paired in the overall redox reaction. To balance a redox equation, the number of electrons in each half-reaction must be made equal by multiplying the appropriate half-reaction by an integer value. The reduction half-reaction above describes the addition of a single electron to the silver cation, but the oxidation half-reaction generates two electrons for each reacting cobalt atom. Thus, the reduction half-reaction is multiplied by two to equalize the number of electrons in the two half-reactions. The two half-reactions are then added together, such that the electrons cancel to give the overall balanced equation with coefficients of two in front of both silver species (Equation 1).

Now, consider Equation 2, the reverse of Equation 1, which describes the potential reaction of silver metal with an aqueous solution of $Co^{2+}$ ions. How can one determine if $Co(s)$ reacts with $Ag^+(aq)$ as noted in Equation 1, or if $Ag(s)$ reacts with $Co^{2+}(aq)$ as in Equation 2? In other words, which reaction will take place?

$$2Ag(s) + Co^{2+}(aq) \rightarrow 2Ag^+(aq) + Co(s) \quad\quad\quad \textbf{Equation 2}$$

The activity series lists cobalt as a more active metal than silver. Therefore, Equation 1 would be the correct prediction. An experiment involving the two sets of reactants in Equations 1 and 2 would, indeed, confirm that prediction to be correct.

## THIS LAB

In this lab, you will be given four cation solutions, each containing $Cu^{2+}(aq)$, $Ag^+(aq)$, $Zn^{2+}(aq)$, or $H^+(aq)$. By exposing five metals, Cu, Zn, Mg, Ag, and Fe to these solutions, a subset of the activity series will be created. Using your observations, you will rank $Mg(s)$, $Fe(s)$, $Ag(s)$, $Zn(s)$, $Cu(s)$, and $H_2(g)$ based on their activity. Balanced redox reactions will then be written for every redox reaction observed. For proper balancing, one should begin with two balanced half-reactions describing the coupled oxidation and reduction by separate equations. The relevant half-reactions, written as oxidations, are

$$Cu \rightarrow Cu^{2+} + 2e^-$$

$$Zn \rightarrow Zn^{2+} + 2e^-$$

$$Ag \rightarrow Ag^+ + e^-$$

$$H_2 \rightarrow 2H^+ + 2e^-$$

$$Mg \rightarrow Mg^{2+} + 2e^-$$

$$Fe \rightarrow Fe^{2+} + 2e^-$$

## *Procedure*

1. Obtain approximately 20 mL of 0.1 $M$ $CuSO_4$, 0.1 $M$ $ZnSO_4$, 0.1 $M$ $AgNO_3$, and 6 $M$ HCl in separate 25-mL Erlenmeyer flasks. Label each flask appropriately and designate a Beral pipette for each solution.

2. Arrange five sets of test tubes (four test tubes per set) in a test tube rack to match the grid shown in the data collection section of this lab. Label each set of test tubes with one of the metals to be used in this laboratory and one of the four ions ($Cu^{2+}$, $Ag^+$, $Zn^{2+}$, and $H^+$) to be added to it. (*Note: You will not need test tubes for $Cu(s)/Cu^{2+}(aq)$, $Zn(s)/Zn^{2+}(aq)$, and $Ag(s)/Ag^+(aq)$ combinations.*) Alternatively, you can place the test tube rack on a sheet of white paper and label your rows and columns of test tubes directly on the paper to match the grid.

3. Add one small piece of metal (wire, ribbon, or lump) to each labeled test tube or column. Wire snips or cutters are recommended for cutting small pieces of metal wire and ribbon.

4. Progressing across the grid, fill each metal-containing test tube no higher than one-third full with each solution as appropriate. Record all observations over a 10-min period on the grid provided in the data collection section. (*Note: Use a different Beral pipette for each different solution.*)

5. Rank $Mg(s)$, $Fe(s)$, $Ag(s)$, $Zn(s)$, $Cu(s)$, and $H_2(g)$ based on their activity.

6. Write balanced net ionic equations for each observed reaction.

## Data Collection

1. Record observations here:

| | Cu²⁺ (aq) | Zn²⁺ (aq) | Ag⁺ (aq) | H⁺ (aq) |
|---|---|---|---|---|
| **Cu(s)** | X | | | |
| **Zn(s)** | | X | | |
| **Ag(s)** | | | X | |
| **Fe(s)** | | | | |
| **Mg(s)** | | | | |

2. Activity ranking, step 5:

_____  _____  _____  _____  _____  _____

**1** (most active)  **2**  **3**  **4**  **5**  **6** (least active)

3. Balanced net ionic equations (based on the results from your grid), step 6:
   (*Note: Phase labels must be included in each equation.*)

## Reaction 1

Relevant Half-Reactions:

    *Oxidation*:

    *Reduction:*

Net Ionic Equation:

## Reaction 2

Relevant Half-Reactions:

    *Oxidation*:

    *Reduction:*

Net Ionic Equation:

## Reaction 3

Relevant Half-Reactions:

    *Oxidation*:

    *Reduction:*

Net Ionic Equation:

## Reaction 4

Relevant Half-Reactions:

    *Oxidation*:

    *Reduction:*

Net Ionic Equation:

## Reaction 5

Relevant Half-Reactions:

    *Oxidation*:

    *Reduction:*

Net Ionic Equation:

## Reaction 6

Relevant Half-Reactions:

*Oxidation*:

*Reduction:*

Net Ionic Equation:

## Reaction 7

Relevant Half-Reactions:

*Oxidation*:

*Reduction:*

Net Ionic Equation:

## Reaction 8

Relevant Half-Reactions:

*Oxidation*:

*Reduction:*

Net Ionic Equation:

## Reaction 9

Relevant Half-Reactions:

*Oxidation*:

*Reduction:*

Net Ionic Equation:

## Reaction 10

Relevant Half-Reactions:

*Oxidation*:

*Reduction:*

Net Ionic Equation:

## Reaction 11

Relevant Half-Reactions:

*Oxidation*:

*Reduction:*

Net Ionic Equation:

## Reaction 12

Relevant Half-Reactions:

*Oxidation*:

*Reduction:*

Net Ionic Equation:

## Reflection Questions

1. For the set of reactions that you studied in this lab, identify the following:

   Best oxidizing agent: _____

   Best reducing agent: _____

   Most easily oxidized species: _____

   Most easily reduced species: _____

2. Based on your activity series:
   (a) Which of the metals used in this lab is *most* likely to be found in the uncombined or metallic state in nature?

   (b) Which metal would be *least* likely to be found uncombined with other elements?

   (c) Explain why copper was a better choice than zinc for use on the Statue of Liberty.

   (d) Is there a metal from your set that would be even better than copper for the Statue of Liberty? If so, which and why? Would this be a practical choice?

3. Write a balanced *molecular* equation describing one of your redox reactions. Identify any spectator ions.

4. Because of its cost and strength, steel components are commonly used in the construction of recreational boats and, to an even greater degree, large ships. As steel is predominately comprised of iron metal, it is prone to corrosion or oxidation over time, especially in salt water. The high concentration of ions in salt water provides a conducting medium that facilitates electron transfer reactions. To combat the breakdown of iron-containing parts, oceangoing boats typically have pieces of zinc metal in contact (either directly or through wire) with the steel components.
   (a) Considering your data on the relative activity of zinc and iron metal, how does the zinc metal protect the steel components from oxidation?

   (b) From a maintenance standpoint, does the boat owner have to periodically monitor the zinc metal? In other words, what happens to the zinc metal over time?

5. The activity of aluminum metal is in between that of magnesium and zinc.
   (a) Would you consider aluminum to be an active, or easily oxidized, metal? Explain.

   (b) Is your answer to part (a) consistent with the apparent stability of aluminum foil in air? Explain.

## Connection

Based on your experience in this lab, draw a connection to something in your everyday life or the world around you (something not mentioned in the background section):

# 12

# Calorimetry

## Objectives

- Construct and utilize a coffee-cup calorimeter to measure heat changes.
- Determine the heat capacity of a calorimeter.
- Determine the specific heat of an unknown metal sample.
- Use Hess's law to experimentally determine the $\Delta H_f^\circ$ of magnesium oxide.

## Materials Needed

- 2 Styrofoam coffee cups and 1 lid
- Thermometer
- Magnetic stir bar
- Magnetic stirring plate
- Hotplate
- 100-mL graduated cylinder
- 600-mL beaker
- 50-mL Erlenmeyer flask
- Small test tube
- Ring stand and test tube clamp
- Burette clamp
- Small piece of unknown metal (provided by instructor)
- 0.15 g magnesium/magnesium tape
- 50 mL of 1.0 $M$ hydrochloric acid (HCl)
- 0.25 g magnesium oxide (MgO)
- Distilled water

## Safety Precautions

- Gloves and safety goggles/glasses should be worn during this lab.
- Never pick up a heated piece of metal with your bare hands to avoid burns.
- Hydrochloric acid (HCl) is toxic by ingestion and inhalation and corrosive to skin and eyes; avoid contact with body tissues.
- Dispose of waste materials as indicated by your instructor.
- In Part III of this lab, hydrogen gas will be generated; as hydrogen is flammable, keep all heat and flames away from your reaction vessel.

# Background

There are two reasons to study or use a chemical reaction: (1) the generation of *products* that have value or (2) the extraction of energy. Chemical changes (and physical changes) are always accompanied by a change in energy. Heat (symbolized by $q$) is energy that either leaves a system (the reaction or process being studied) into the surroundings, or else enters from the surroundings into the system. Heat flows from a higher-temperature domain (either the system or surroundings) to one with a lower temperature until the temperature in both domains is equal (thermal equilibrium). The heat change in the system is equal in magnitude, but necessarily opposite in sign to that for the surroundings:

$$q_{system} = -(q_{surroundings}) \qquad \qquad \textbf{Equation 1}$$

When heat flows into the system (i.e., heat is absorbed), $q > 0$ and the process is *endothermic*. In contrast, when heat flows out of the system (i.e., heat is released), $q < 0$ and the process is *exothermic*. If constant pressure conditions are used (e.g., a reaction on a lab bench under atmospheric pressure), then $q_P = \Delta H$, which stands for the *enthalpy* change of the system. The enthalpy change for a given process depends only on the initial and final states of the process, and not on the path that the process takes from initial to final state (i.e., $\Delta H$ is a state function). Thus, for a chemical reaction:

$$\Delta H_{rxn} = H_{products} - H_{reactants} \qquad \begin{array}{ll} \Delta H > 0 & \text{endothermic} \\ \Delta H < 0 & \text{exothermic} \end{array} \qquad \textbf{Equation 2}$$

A balanced chemical equation (with the physical states of all reactants and products included) accompanied by the heat of reaction ($\Delta H_{rxn}$) is known collectively as a *thermochemical equation*. The thermochemical equation describes the stoichiometric relationship between the mass/mole conversions of the reaction and the energy changes associated with that reaction. For example,

$$H_2O(s) \rightarrow H_2O(l) \qquad \Delta H = +6.01 \text{ kJ mol}^{-1}$$

This thermochemical equation indicates that 6.01 kJ of heat are absorbed from the surroundings (an endothermic process) during the conversion of 1 mole of ice to 1 mole of liquid water.

## CALCULATION OF $\Delta H$ USING HESS'S LAW

The enthalpy change, $\Delta H$, that occurs when reactants are converted to products is the same whether the reaction occurs in one step or a series of steps (i.e., enthalpy is a state function). If a reaction is carried out in a series of steps, the sum of the enthalpies for each step will equal the enthalpy change for the total reaction. In other words, if two chemical equations can be summed to give a third equation, the values of $\Delta H$ for the two equations can be combined in the same manner to give $\Delta H$ for the third equation—this is known as *Hess's law*. In order to use Hess's law, it is first necessary to obtain reaction enthalpies for reactions that can be summed together appropriately. There are two simple guidelines for manipulating thermochemical equations when attempting to combine them using Hess's law:

1. Multiplication of an equation by a numerical factor gives a $\Delta H$ value equal to the original value multiplied by the same factor ($\Delta H$ is an extensive property—it depends on the amount of material involved):

$$C(s) + O_2(g) \rightarrow CO_2(g) \qquad \Delta H = -393.5 \text{ kJ mol}^{-1}$$
$$3C(s) + 3O_2(g) \rightarrow 3CO_2(g) \qquad \Delta H = -1181 \text{ kJ mol}^{-1}$$

2. Reversal of an equation changes the sign but not the magnitude of $\Delta H$:

$$C(s) + O_2(g) \rightarrow CO_2(g) \qquad \Delta H = -393.5 \text{ kJ mol}^{-1}$$
$$CO_2(g) \rightarrow C(s) + O_2(g) \qquad \Delta H = +393.5 \text{ kJ mol}^{-1}$$

# STANDARD ENTHALPIES OF FORMATION

The standard heat of formation ($\Delta H_f^\circ$) is the heat change that results when exactly *1 mole* of a compound is formed from its *elements* in their *standard states*. The standard state is the most stable state of the element found at 1 atm (and typically at 25°C). Examples of standard states are $O_2(g)$ for oxygen and graphite for carbon. By convention, $\Delta H_f^\circ$ for an element in its standard state is 0. Examples of formation reactions follow:

$$\tfrac{1}{2}N_2(g) + \tfrac{3}{2}H_2(g) \rightarrow NH_3(g) \qquad \Delta H_f^\circ\,[NH_3(g)] = -46.3 \text{ kJ mol}^{-1}$$
$$2Na(s) + C\,(graphite) + \tfrac{3}{2}O_2(g) \rightarrow Na_2CO_3(s) \qquad \Delta H_f^\circ\,[Na_2CO_3(s)] = -1130.9 \text{ kJ mol}^{-1}$$

$\Delta H_f^\circ$ values can then be used to calculate the standard enthalpy change for any chemical reaction:

$$\Delta H_{rxn}^\circ \text{ (standard enthalpy of reaction)} = \Delta H_{products} - \Delta H_{reactants} \qquad \textbf{Equation 3}$$

$$\Delta H_{rxn}^\circ = \sum [n \times \Delta H_f^\circ]_{(products)} - \sum [m \times \Delta H_f^\circ]_{(reactants)} \qquad \textbf{Equation 4}$$

where "$\Sigma$" = sum of the $\Delta H_f^\circ$ values, and "$n$" and "$m$" = coefficients from the balanced chemical reaction.


# MEASUREMENT OF $\Delta H$ USING CALORIMETRY

Calorimetry is a standard technique used to experimentally measure the heat change associated with any process. A calorimeter is an insulated device that prevents (or at least significantly inhibits) the flow of heat between its interior and the surroundings, allowing for the measurement of $\Delta H$ (if it is under conditions of constant pressure). The basis for the determination of $\Delta H$ is that the heat released by a chemical reaction or physical change (i.e., the system) is entirely absorbed by the calorimeter (the surroundings) and vice versa (heat absorbed by the system equals heat lost by the surroundings). Further, the heat exchange will occur until both the system and surroundings are at the same temperature (thermal equilibrium). Thus, in a typical calorimetry experiment the temperature of the surroundings is recorded and reflects the heat that either comes from or goes into the system.

$$q_{object\,1} = -q_{object\,2} \qquad \textbf{Equation 5}$$

$$T_{final\,(object\,1)} = T_{final(object\,2)} \qquad \textbf{Equation 6}$$

The relationship between the temperature change of a substance and heat is given by

$$q = s \times m \times \Delta T \qquad \textbf{Equation 7}$$

where $m$ is the mass of the substance, $\Delta T$ is the change in temperature (determines sign of $q$), and $s$ is the *specific heat* of the substance, an intensive property that specifies the amount of heat required to raise the temperature of 1 g of a substance by 1°C (units = $J\,g^{-1}\,°C^{-1}$). Different substances have different specific heats; some common examples are shown.

| Substance | Specific Heat ($J\,g^{-1}\,°C^{-1}$) |
|---|---|
| $H_2O(l)$ | 4.184 |
| $Al(s)$ | 0.900 |
| $C(s, graphite)$ | 0.720 |
| $Fe(s)$ | 0.444 |
| $Hg(l)$ | 0.139 |

The larger the specific heat, the more energy is required to cause a temperature change. So, for example, one might want the handle of a cooking pot to have a much higher specific heat than the pot itself to allow both for efficient heating of the pot and its ease of handling.

Combination of the specific heat and mass of an object gives heat capacity ($C$), or the amount of heat required to raise the temperature of an object by 1°C (units = J °C$^{-1}$).

$$C = s \times m \qquad \textbf{Equation 8}$$

Importantly, by knowing the heat capacity of the calorimeter and measuring its $\Delta T$, the heat ($q$) absorbed or released by a system can be determined. Note that in a calorimeter, we have a collection of substances (each with a different specific heat) that function together to absorb or release heat with respect to the system. For today's lab, the calorimeter and water will make up the surroundings.

$$q = C \times \Delta T \qquad \textbf{Equation 9}$$
$$q_{system} = -[q_{water} + q_{calorimeter}] \qquad \textbf{Equation 10}$$

A negative sign is incorporated on the right side of Equation 10 to indicate that heat lost = heat gained. For example, an exothermic reaction will give a positive $\Delta T$ for the water and walls of the calorimeter as heat is absorbed by the surroundings. To get $q_{calorimeter}$, the calorimeter must be *calibrated* (i.e., determine $\Delta T$ for the calorimeter when a known reaction is carried out in the calorimeter) to determine its actual heat capacity.

For this lab, you will be building and using a coffee-cup calorimeter (see Figure 2 in Procedure section). This type of calorimeter is good for measuring heat changes under constant pressure conditions (thus, $\Delta H$ values) for processes such as neutralization reactions, dissolving solids (solution process), and dilutions as well as for determining the specific heat or heat capacity of objects. The system is defined as the object placed in the calorimeter or the chemicals dissolved in the water; the surroundings are the water and the walls of the calorimeter. An increase in the water temperature indicates that heat is released by the system, thus an exothermic process and $\Delta H < 0$, while a decrease in water temperature indicates an endothermic process with $\Delta H > 0$.

## CORRECT DETERMINATION OF $\Delta T$

Critical to calorimetry is an ability to accurately measure $\Delta T$. This is affected by the speed with which a reaction occurs and how well the calorimeter is insulated. A plot of temperature vs. time for an exothermic reaction under three different scenarios illustrates this point nicely (Figure 1):

1. Instantaneous reaction in a perfectly insulated calorimeter (no heat loss)
2. Instantaneous reaction in a calorimeter that is not completely insulated (allows heat loss)
3. Noninstantaneous reaction in a calorimeter that is not perfectly insulated (most realistic)

In situation 1, all heat is contained, so the temperature increases and stays constant; in this case, we can simply look at $T_{final} - T_{initial}$ to get $\Delta T$. For situation 2, the temperature decreases as the calorimeter loses heat to the surroundings; in this instance, we would need to use the maximum temperature as $T_{final}$. Finally in situation 3, the theoretical maximum temperature (if all heat had been contained in calorimeter) is never reached since the heat is leaching from the calorimeter as the reaction nears completion; for this case, one needs to do a linear extrapolation of the gradual cooling part of the graph back to time = 0 to get an estimation of the maximum temperature, which we assign as $T_{final}$.

## THIS LAB

For this lab there will be three specific tasks, which are outlined below.

1. *Build and calibrate a coffee-cup calorimeter:* You will calibrate your calorimeter by adding a known mass of warm water to a known mass of cold water in the calorimeter and then determining the temperature change. Because you know the specific heat of water, you can determine how much heat is absorbed by the calorimeter.

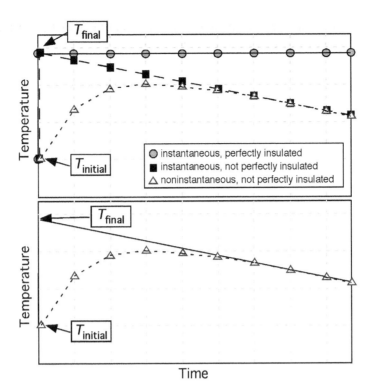

**Figure 1** (Top) Temperature vs. time plots for an exothermic reaction under three scenarios: instantaneous reaction in a perfectly insulated calorimeter (gray circles), instantaneous reaction in a calorimeter that leaks heat (black squares), and a noninstantaneous reaction in a not perfectly insulated calorimeter (open triangles). (Bottom) Plot for third scenario showing extrapolation to time = 0 to get $T_{final}$.

2. *Use your calorimeter to determine the specific heat of an unknown metal, and from that determine the identity of the metal:* To measure the specific heat of a substance (e.g., a metal), a known mass of the metal is heated to a measured temperature, and then added to a known mass of water of known temperature (cooler) in the calorimeter. The metal cools and the water increases in temperature as heat is transferred, with both eventually reaching the same temperature. From this, the specific heat of the metal can be determined.

3. *Use the calorimeter and Hess's law to experimentally determine the heat of formation of MgO:* Many metallic elements are not found in nature as pure metals, but instead in oxidized form as parts of compounds such as metal oxides; common examples include $SiO_2$, iron oxides, and $Al_2O_3$. Some metal oxides form a thin protective layer on the surface of the metal, which prevents it from oxidizing further ($Al_2O_3$ and CrO), while other oxides are more porous and provide very little protection (e.g., iron rusts completely over time to give $Fe_2O_3$). Knowing the enthalpy of formation of metal oxides is therefore important, but can be difficult to determine. However, by using Hess's law, we can get these values by instead measuring $\Delta H$ for related reactions. For this lab, you will be determining $\Delta H_f^\circ$ for MgO:

1. $Mg(s) + 2HCl(aq) \rightarrow MgCl_2(aq) + H_2(g, 1\ atm)$     $\Delta H_1$     (measure)

2. $MgO(s) + 2HCl(aq) \rightarrow MgCl_2(aq) + H_2O(l)$     $\Delta H_2$     (measure)

3. $H_2(g, 1\ atm) + \frac{1}{2}O_2(g, 1\ atm) \rightarrow H_2O(l)$     $\Delta H_3 = -285.840\ kJ\ mol^{-1}$
                                                                              (from $\Delta H_f$)

_____

$Mg(s) + \frac{1}{2}O_2(g, 1\ atm) \rightarrow MgO(g, 1\ atm)$     $\Delta H_{total} = \Delta H_f^\circ\ [MgO(s)]$
                                                                                        (what we want)

In this experiment, you will use excess HCl(*aq*) in steps 1 and 2, as both the source of the reactant and a component of the surroundings. We will assume that the density of 1.0 *M* HCl is equivalent to the density of water (1 g mL$^{-1}$), and that the specific heat of 1.0 *M* HCl is equivalent to the specific heat of water (4.184 J g$^{-1}$ °C$^{-1}$).

## Procedure

### I. Construct and Calibrate Coffee-Cup Calorimeter

1. Set up a hot water bath by putting ~500 mL of water into a 600-mL beaker; place the beaker on a hotplate and set on high. Pay attention as additional water may need to be added if the water level gets too low. [NOTE: if stirrer/hotplates are in limited supply in the lab, multiple students/teams might consider sharing a larger hot water bath.]
2. Take two coffee cups, nest one inside the other, and determine their combined mass.
3. Add ~70 mL of room-temperature deionized water ("cool water") to the nested coffee cups and determine the total mass.
4. Construct the coffee-cup calorimeter (Figure 2):
   - Use two nested Styrofoam coffee cups.
   - Add a magnetic stir bar.
   - Add a single polystyrene lid with 1 small hole (for the thermometer).
   - Place the calorimeter on a magnetic stirring plate.

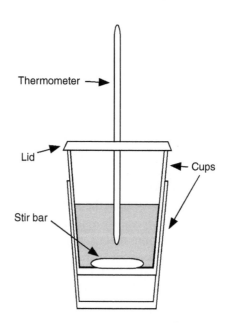

**Figure 2** **Diagram showing assembly of a coffee-cup calorimeter.**

5. Take a 50-mL Erlenmeyer flask; add ~30 mL of room-temperature deionized water to the flask and record the mass of the water (this will be the "hot water").
6. Suspend the Erlenmeyer flask in the water bath, making sure the water in the flask is below the water level in the bath; allow the flask to sit in the water bath for ~10 to 15 min.
7. Use your thermometer to record the temperature of the water in the water bath beaker—this will be the temperature of the hot water.
8. Before adding the hot water to the calorimeter, insert your thermometer through the hole in the calorimeter lid and use a clamp to secure it—be sure that the bulb of the thermometer is positioned approximately in the middle of the water (i.e., not touching the bottom or sides of the cup); record the temperature (this will be the temperature of the cool water and should be recorded both in Line #5 in the Data Acquisition section, as well as for time = 0 temperature in #6).

9. After you are sure the thermometer is high enough to not contact the stir bar, turn on the magnetic stirrer; be sure that there is no splashing.

10. Now, use the clamp to remove the Erlenmeyer flask from the bath; lift the calorimeter lid and carefully (but quickly) pour the hot water into the calorimeter—this is time = 0 sec. Replace the lid and be sure that the thermometer bulb is in the water.

11. Record the time and temperature every 15 to 30 sec for a total period of 8 min. The time between recordings can be increased as the temperature changes become small.

12. Once done, empty the calorimeter and dry completely being careful not to crack the cups or lid.

## II. Measure Specific Heat of an Unknown Metal

13. If necessary, add more hot water to the water bath and keep heating until boiling.

14. Add ~40 mL of room-temperature of deionized water to the nested coffee cups and determine the total mass (cups + water); then assemble the calorimeter (add stir bar and lid).

15. Determine the mass of a small piece of unknown metal (dry); place the metal in a small test tube and clamp it in the water bath (the level of the metal should be below the water bath level). The metal should stay in the water bath for ~5 to 10 min to come to appropriate temperature.

16. Use a thermometer to record the temperature of the boiling water (this will also be the initial temperature of the metal); remove and dry the thermometer.

17. Insert the thermometer into your calorimeter and carefully clamp it into place, being sure that the bulb of the thermometer is positioned approximately in the middle of the water (i.e., not touching the bottom or sides of the cup); record the temperature (lid should be closed; this will be the initial temperature of the water recorded in Line #10, as well as for time = 0 temperature in #11); leave the thermometer in the calorimeter.

18. Remove the test tube from the boiling water; quickly dry the outside, lift the lid of the calorimeter and carefully pour the metal into the calorimeter (don't splash—i.e., keep all water in cup); replace the lid.

19. Stir the water in the calorimeter; record the time and temperature every 15 to 30 sec for a total period of 8 min (it should increase as the metal transfers heat to the water).

20. Remove the metal from the calorimeter and pour out the water; dry the metal sample thoroughly.

21. Repeat steps 13 through 20 to get a second set of readings (no need to determine the mass of the metal sample again).

## III. Hess's Law

*Magnesium Reaction*

22. Add ~25 mL of 1.0 $M$ HCl to the two nested Styrofoam cups and determine the total mass (cups + HCl); then assemble the calorimeter (add stir bar and lid).

23. Insert the thermometer in the calorimeter and clamp in place; be sure the thermometer bulb is immersed in the liquid and not touching the sides or bottom of the cup, and is safely above the stir bar. Record the temperature of HCl (this will be the initial temperature of the HCl recorded in Line #13, as well as for time = 0 temperature in Line #15). Begin stirring the HCl.

24. Measure approximately 0.15 g of Mg metal and record mass precisely.

25. Carefully add the Mg to the acid—be careful to avoid splashing. Replace the lid.

26. Record the time and temperature every 15 to 30 sec for a total period of 8 min; be sure the reaction has gone to completion (i.e., the limiting reagent, Mg, has completely reacted)

27. Once finished, dispose of the contents of the calorimeter as indicated by your instructor and rinse with water.

*Magnesium Oxide Reaction*

28. Perform a similar experiment as described in procedures 22 through 27 using 25.0 mL of 1.0 $M$ HCl and 0.25 g magnesium oxide (MgO).

## I. Construct and Calibrate Coffee-Cup Calorimeter

**1.** Mass of two Styrofoam cups: _____

**2.** Mass of cups + ~70 mL "cool water" (i.e., room-temperature water): _____

**3.** Mass of ~30 mL "hot water", $mass_{hot}$: _____

**4.** Temperature of hot water (i.e., water in water bath), $T_{i, hot}$: _____

**5.** Temperature of cool water, $T_{i, cool}$: _____

**6.** Temperature vs. time data:

| Time | Temperature (°C) | | Time | Temperature (°C) |
|------|------------------|--|------|------------------|
| 0 | | | | |
| | | | | |
| | | | | |
| | | | | |
| | | | | |
| | | | | |
| | | | | |
| | | | | |
| | | | | |
| | | | | |
| | | | | |
| | | | | |
| | | | | |
| | | | | |
| | | | | |
| | | | | |
| | | | | |

## II. Measure Specific Heat of an Unknown Metal

|  | Trial 1 | Trial 2 |
|--|---------|---------|
| **7.** Mass of cups + ~40 mL room-temperature water: | _____ | _____ |
| **8.** Mass of unknown metal, $mass_{metal}$: | _____ | same as trial 1 |
| **9.** Temperature of boiling water (same as initial temp of metal, $T_{i\ metal}$): | _____ | _____ |
| **10.** Temperature of water in calorimeter, $T_{i\ water}$: | _____ | _____ |

**11.** Temperature vs. time data:

## Trial 1

| Time | Temperature (°C) | | Time | Temperature (°C) |
|---|---|---|---|---|
| 0 | | | | |
| | | | | |
| | | | | |
| | | | | |
| | | | | |
| | | | | |
| | | | | |
| | | | | |
| | | | | |
| | | | | |
| | | | | |
| | | | | |
| | | | | |
| | | | | |
| | | | | |
| | | | | |
| | | | | |
| | | | | |

## Trial 2

| Time | Temperature (°C) | | Time | Temperature (°C) |
|---|---|---|---|---|
| 0 | | | | |
| | | | | |
| | | | | |
| | | | | |
| | | | | |
| | | | | |
| | | | | |
| | | | | |
| | | | | |
| | | | | |
| | | | | |
| | | | | |
| | | | | |
| | | | | |
| | | | | |
| | | | | |
| | | | | |
| | | | | |

### III. Hess's Law

*Magnesium Reaction*

**12.** Mass of cups + ~25 mL HCl: _____

**13.** Temperature of HCl in calorimeter, $T_{i\ HCl}$: _____

**14.** Mass of Mg metal: _____

**15.** Temperature vs. time data:

| Time | Temperature (°C) | | Time | Temperature (°C) |
|---|---|---|---|---|
| 0 | | | | |
| | | | | |
| | | | | |
| | | | | |
| | | | | |
| | | | | |
| | | | | |
| | | | | |
| | | | | |
| | | | | |
| | | | | |
| | | | | |
| | | | | |
| | | | | |
| | | | | |
| | | | | |

### Magnesium Oxide Reaction

**16.** Mass of cups + ~25 mL HCl: _____

**17.** Temperature of HCl in calorimeter, $T_{i\ HCl}$: _____

**18.** Mass of MgO: _____

**19.** Temperature vs. time data:

| Time | Temperature (°C) | | Time | Temperature (°C) |
|------|------------------|---|------|------------------|
| 0 | | | | |
| | | | | |
| | | | | |
| | | | | |
| | | | | |
| | | | | |
| | | | | |
| | | | | |
| | | | | |
| | | | | |
| | | | | |
| | | | | |
| | | | | |
| | | | | |
| | | | | |
| | | | | |
| | | | | |

## *Analysis*

### I.   Construct and Calibrate Coffee-Cup Calorimeter

**20.** Mass of cool water: $mass_{cool} = mass_{cups + cool water} - mass_{cups}$

$$= (\#2) - (\#1) \qquad\qquad = \rule{4cm}{0.4pt}$$

**21.** Maximum temperature: construct a plot of temperature vs. time using data from #6, similar to that shown in Figure 1. From that, determine the maximum temperature (i.e., $T_f$)—this will be the y-intercept of the extrapolation line; the line can be obtained using Excel, a graphing calculator, or the least-squares method you utilized in Laboratory 4.

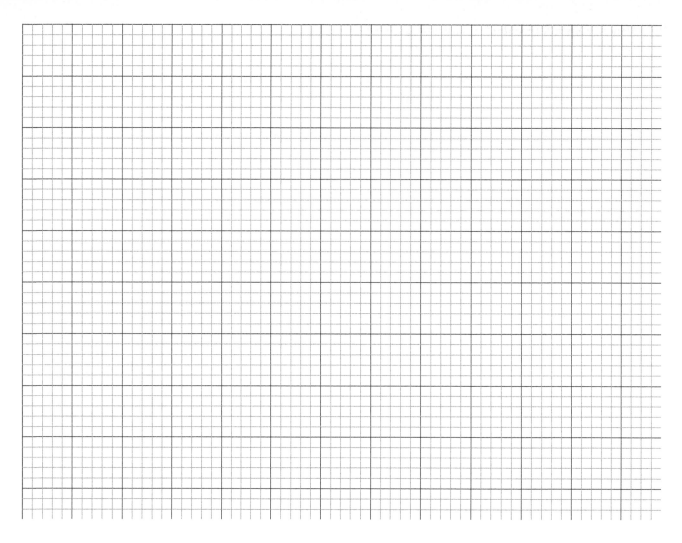

$$T_f = \text{_____}$$

**22.** Heat exchanged between the hot and cool water:

$$-q_{\text{hot water}} = q_{\text{cool water}} + q_{\text{calorimeter}} \qquad \textbf{Equation 11}$$

a. $q_{\text{hot water}} = (s_{\text{water}}) \times (\text{mass}_{\text{hot}}) \times \Delta T$
   $$= (4.184 \text{ J g}^{-1}\,^{\circ}\text{C}^{-1}) \times (\#3) \times [(\#21) - (\#4)] \qquad = \text{_____}$$

b. $q_{\text{cool water}} = (s_{\text{water}}) \times (\text{mass}_{\text{cool}}) \times \Delta T$
   $$= (4.184 \text{ J g}^{-1}\,^{\circ}\text{C}^{-1}) \times (\#20) \times [(\#21) - (\#5)] \qquad = \text{_____}$$

c. $q_{\text{calorimeter}} = -(q_{\text{hot water}} + q_{\text{cool water}})$
   $$= -[(\#22a) + (\#22b)] \qquad = \text{_____}$$

**23.** Total heat capacity of calorimeter:

$q_{\text{calorimeter}} = (C_{\text{calorimeter}}) \times \Delta T$
$C_{\text{calorimeter}} = (\#22c) \, / \, [(\#21) - (\#5)] \qquad = \text{_____}$

## II. Measure Specific Heat of an Unknown Metal

**24.** Mass of water: $\text{mass}_{\text{water}} = \text{mass}_{\text{cups + water}} - \text{mass}_{\text{cups}}$

$$= (\#7) - (\#1) \qquad = \underline{\hspace{3cm}} \qquad \underline{\hspace{3cm}}$$

$$\text{trial 1} \qquad\qquad\qquad \text{trial 2}$$

**25.** Maximum temperature of water: construct a plot of temperature vs. time using data from #11 *for each trial*, similar to that shown in Figure 1. From that, determine the maximum temperature (i.e., $T_f$)—this will be the *y*-intercept of the extrapolation line; the line can be obtained using Excel, a graphing calculator, or the least-squares method you utilized in Laboratory 4.

**Trial 1:**

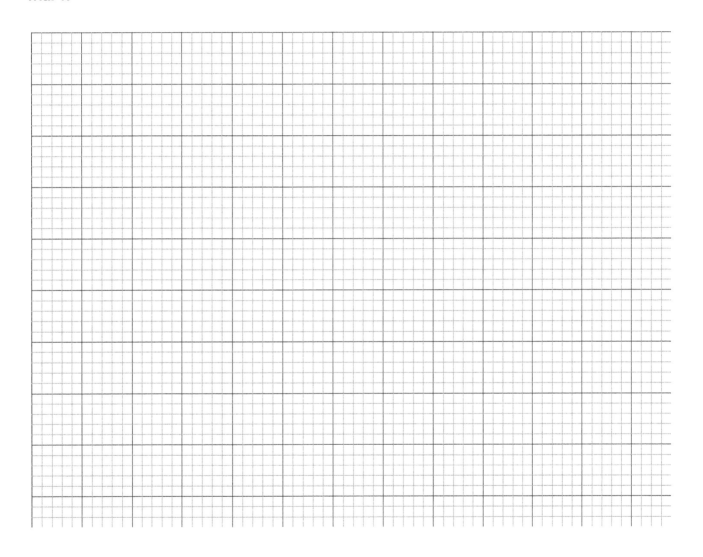

$$T_f = \underline{\hspace{4cm}} \text{(trial 1)}$$

**Trial 2:**

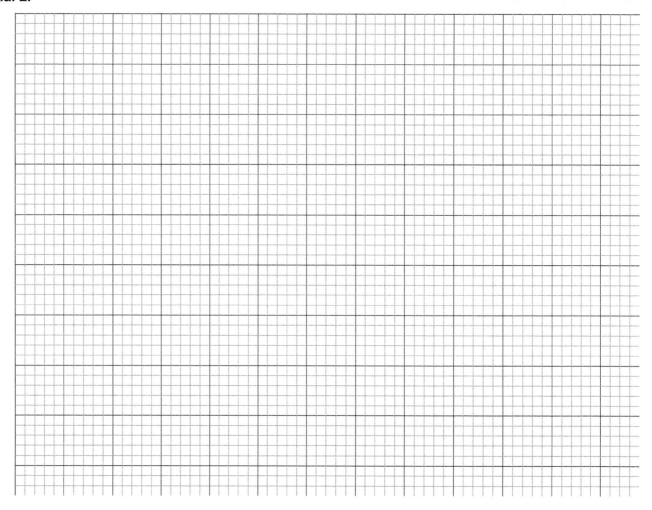

$T_f =$ _____ (trial 2)

**26.** Heat exchanged between the metal and water:

$$-q_{metal} = q_{water} + q_{calorimeter}$$    **Equation 12**

a.  $q_{water} = (s_{water}) \times (mass_{water}) \times (T_f - T_{i\,water})$
    $= (4.184\ \text{J g}^{-1}\,°\text{C}^{-1}) \times (\#24) \times [(\#25) - (\#10)]$    $=$ _____    _____

    trial 1          trial 2

b.  $q_{calorimeter} = (C_{calorimeter}) \times (T_f - T_{i\,water})$
    $= (\#23) \times [(\#25) - (\#10)]$    $=$ _____    _____

    trial 1          trial 2

c.  $q_{metal} = -(q_{water} + q_{calorimeter})$
    $= -[(\#26a) + (\#26b)]$    $=$ _____    _____

    trial 1          trial 2

**27.** Specific heat of unknown metal:

$$q_{metal} = (s_{metal}) \times (mass_{metal}) \times \Delta T$$

$$s_{metal} = (\#26c) / (\#8) \times [(\#25) - (\#9)]$$    $=$ _____    _____

    trial 1          trial 2

$s_{metal}(\text{average}) =$ _____

### III. Hess's Law

**Magnesium Reaction:**     1.    $Mg(s) + 2HCl(aq) \rightarrow MgCl_2(aq) + H_2(g, 1\ atm)$          $\Delta H_1$

$$\Delta H_1 = q_1 = q_{HCl} + q_{calorimeter}$$          **Equation 13**

**28.** Mass of HCl:    $mass_{HCl} = mass_{cups\ +\ HCl} - mass_{cups}$

$= (\#12) - (\#1)$          $=$ _____

**29.** Maximum temperature of HCl: construct a plot of temperature vs. time using data from #15, similar to that shown in Figure 1. From that, determine the maximum temperature (i.e., $T_f$)—this will be the y-intercept of the extrapolation line; the line can be obtained using Excel, a graphing calculator, or the least-squares method you utilized in Laboratory 4.

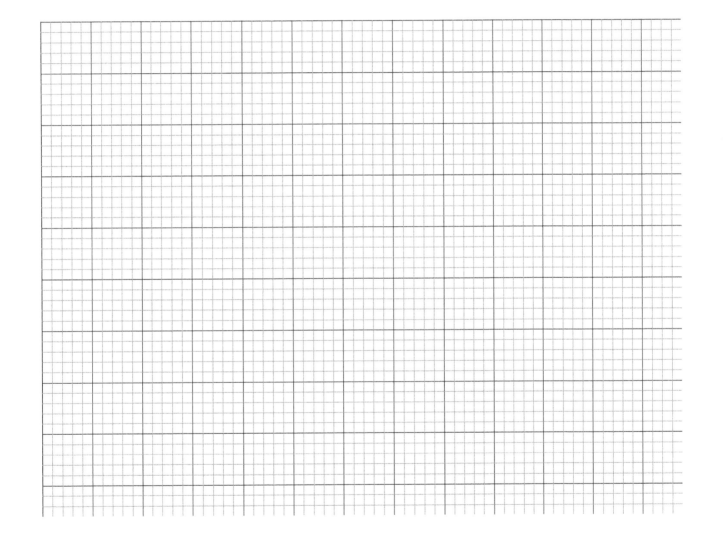

$T_f =$ _____

**30.** Heat of reaction 1, $q_1$:

a.  $q_{HCl} = (s_{HCl}) \times (mass_{HCl}) \times (T_f - T_{i\ HCl})$

$= (4.184\ J\ g^{-1}\ °C^{-1}) \times (\#28) \times [(\#29) - (\#13)]$          $=$ _____

b. $q_{calorimeter} = (C_{calorimeter}) \times (T_f - T_{i\,HCl})$

   $= (\#23) \times [(\#29) - (\#13)]$ $\qquad = $ _____

c. $q_1 = (\#30a) + (\#30b)$ $\qquad = $ _____

**31.** Moles of Mg = (mass of Mg) / (atomic mass of Mg)

   $= (\#14) / (24.31 \text{ g mol}^{-1})$ $\qquad = $ _____

**32.** Enthalpy of reaction 1, $\Delta H_1$: $\qquad \Delta H_1 = q_1 / (\text{mol Mg})$

   $= (\#30c) / (\#31)$ $\qquad = $ _____

**Magnesium Oxide Reaction:** $\qquad$ 2. $\quad MgO(s) + 2HCl(aq) \rightarrow MgCl_2(aq) + H_2O(l)$ $\qquad \Delta H_2$

**33.** Mass of HCl: $mass_{HCl} = mass_{cups + HCl} - mass_{cups}$

   $= (\#16) - (\#1)$ $\qquad = $ _____

**34.** Maximum temperature of HCl: construct a plot of temperature vs. time using data from #19, similar to that shown in Figure 1. From that, determine the maximum temperature (i.e., $T_f$)—this will be the y-intercept of the extrapolation line; the line can be obtained using Excel, a graphing calculator, or the least-squares method you utilized in Laboratory 4.

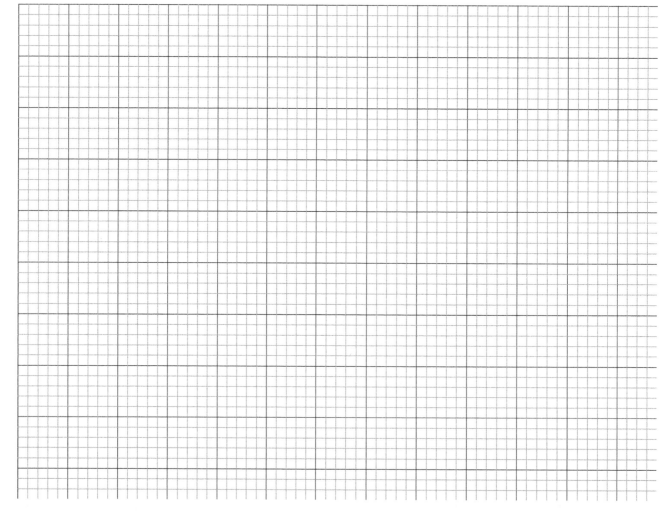

$\qquad\qquad\qquad\qquad\qquad\qquad\qquad T_f = $ _____

**35.** Heat of reaction 2, $q_2$:

a. $q_{HCl} = (s_{HCl}) \times (\text{mass}_{HCl}) \times (T_f - T_{i\,HCl})$

$= (4.184 \text{ J g}^{-1}\,°C^{-1}) \times (\#33) \times [(\#34) - (\#17)]$     = _____

b. $q_{\text{calorimeter}} = (C_{\text{calorimeter}}) \times (T_f - T_{i\,HCl})$

$= (\#23) \times [(\#34) - (\#17)]$     = _____

c. $q_2 = (\#35a) + (\#35b)$     = _____

**36.** Moles of MgO = (mass of MgO) / (formula mass of MgO)

$= (\#18) / (40.31 \text{ g mol}^{-1})$     = _____

**37.** Enthalpy of reaction 2, $\Delta H_2$:     $\Delta H_2 = q_2 / (\text{mol MgO})$

$= (\#35c) / (\#36)$     = _____

**38.** Using the following thermochemical equations and Hess's law:

1. $Mg(s) + 2HCl(aq) \rightarrow MgCl_2(aq) + H_2(g, 1 \text{ atm})$     $\Delta H_1$

2. $MgO(s) + 2HCl(aq) \rightarrow MgCl_2(aq) + H_2O(l)$     $\Delta H_2$

3. $H_2(g, 1 \text{ atm}) + \frac{1}{2}O_2(g, 1 \text{ atm}) \rightarrow H_2O(l)$     $\Delta H_3 = -285.840 \text{ kJ mol}^{-1}$

determine the $\Delta H_f°$ for MgO:

$Mg(s) + \frac{1}{2}O_2(g, 1 \text{ atm}) \rightarrow MgO(g, 1 \text{ atm})$     $\Delta H_f°\,[MgO(s)] =$ _____

## Reflection Questions

1. Based on the temperature vs. time plots for the various parts of this lab, is your coffee-cup calorimeter well insulated? If not, suggest ways that the calorimeter might be improved.

2. If you had ignored the heat absorbed by the calorimeter (Part I), what would be the impact to your results in Parts II and III?

3. If the maximum *recorded* temperature is used in Part II, rather than the extrapolated temperature, will the measured specific heat capacity of the metal be higher or lower than the real value? Explain.

4. Given the specific heats for the following metals, identify which metal is the most likely identity of your unknown metal, and determine the % error for your measured specific heat. Explain possible reasons for any discrepancies.

| Metal | Specific Heat (J g$^{-1}$ °C$^{-1}$) | Metal | Specific Heat (J g$^{-1}$ °C$^{-1}$) |
|---|---|---|---|
| Ag | 0.235 | Fe or Ni | 0.44 |
| Al | 0.90 | Mg | 1.02 |
| Ba | 0.15 | Pb or W | 0.13 |
| Bi | 0.11 | Sr | 0.30 |
| Ca | 0.63 | Zn | 0.39 |
| Cd or Sn | 0.23 | | |

5. Look up the value for the heat of formation of MgO($s$) and calculate your % error. Explain possible reasons for any discrepancies.

6. Why would it be difficult to study the enthalpy of formation of MgO($s$) by directly using your coffee-cup calorimeter (i.e., why is Hess's law a good way to study this reaction?)?

## Connection

Based on your experience in this lab, draw a connection to something in your everyday life or the world around you (something not mentioned in the background section):

# 13

# Gas Laws: Boyle's Law and Experimental Determination of the Ideal Gas Constant

## Objectives

- Verify the relationship between pressure and volume for a gas (Boyle's law).
- Use a chemical reaction and the ideal gas law to experimentally determine the ideal gas law constant $R$.

## Materials Needed

- Tygon tubing (at least 48 in long)—diameter should fit firmly over the end of a 50-mL plastic syringe
- Tape or clamps to hold tubing in a "U"
- Graph paper
- Water in a squirt bottle
- 50-mL plastic syringe
- Plastic ruler (metric markings)
- 100-mL graduated cylinder
- 600-mL beaker
- 18 mm $\times$ 150 mm test tube
- Thermometer
- Ring stand and clamp to hold test tube
- Rubber stopper (fits in test tube) with single hole containing short piece of glass tubing
- Piece of Tygon tubing or rubber hose
- Piece of glass tubing with "U"-shaped hook on one end
- ~0.200 to 0.225 g Zn metal
- 6.0 $M$ HCl (8 mL)
- 1 piece of filter paper (5.5 cm)

## Safety Precautions

- Gloves and safety goggles/glasses should be worn during this lab.
- Hydrochloric acid (HCl) is toxic by ingestion and inhalation and corrosive to skin and eyes; avoid contact with body tissues.

## Background

Gases play a huge role in everyday life. Beyond the simple fact that the atmosphere is a mixture of gases critical for life on this planet, gases are important in numerous other ways:

- *Propellant in aerosols:* hydrochlorofluorocarbons (HCFCs) are compounds that are mixed in liquid form with the product to be discharged (e.g., deodorant) in the can, filling the head space in the container as a gas; when the nozzle is pushed down, the HCFCs force the product out of the nozzle.
- *Fire extinguishers:* $CO_2$ is used to pressurize the container and force water out of the extinguisher.
- *Carbonation of your favorite soft drinks:* during the bottling process, $CO_2$ is added to the head space in the soda container and dissolved with pressure in the liquid.
- *Tire inflation*
- *Hot air balloons:* a hot air balloon floats because the heated air inside the balloon is less dense than that outside, increasing the balloon's buoyancy.

## THE GAS LAWS

When viewed as ideal gases, gas properties are independent of the gas type because the gas particles are far apart and are viewed as noninteracting. The situation is very different for both solids and liquids, condensed states of matter, where the particles are in physical contact with each other. This allows us to define general relationships between temperature ($T$), pressure ($P$), volume ($V$), and number of moles ($n$) that hold for all ideal gases. These relationships are called the *gas laws*.

1. *Boyle's Law:* Robert Boyle, an English philosopher, chemist, physicist, and alchemist, is considered one of the founders of modern chemistry. In the early 1660s, he established a relationship between the pressure and volume of a gas; specifically, the volume of a specific quantity of gas varies inversely to its pressure under constant temperature conditions:[1]

$$V \alpha \frac{1}{P} \qquad\qquad V = k \times \frac{1}{P} \qquad\qquad \textbf{Equation 1}$$

A plot of volume vs. pressure gives the curve in Figure 1 (left), while a plot of volume vs. 1/P yields a straight line.

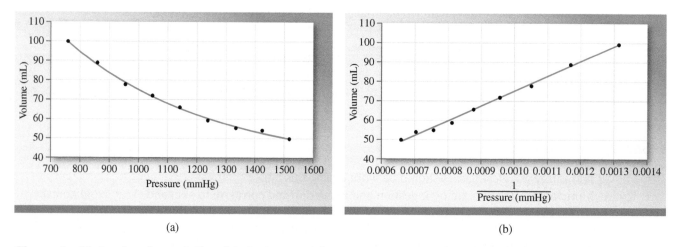

(a)                                                          (b)

**Figure 1   Plots showing relationship between volume and pressure (Boyle's law).**

2. *Charles's/Gay-Lussac's Law:* Charles's law describes the relationship between the volume of a gas and its temperature. This direct proportionality was first established by Jacques Charles, a French scientist in the 1780s, but not published at the time. This was left to Joseph Louis Gay-Lussac in 1802.[2] As can be shown with a simple balloon, heating a gas causes expansion and cooling a gas, contraction.

$$V \, \alpha \, T \qquad\qquad V = k' \times T \qquad\qquad \textbf{Equation 2}$$

A plot of volume vs. temperature yields a straight line (Figure 2, left).

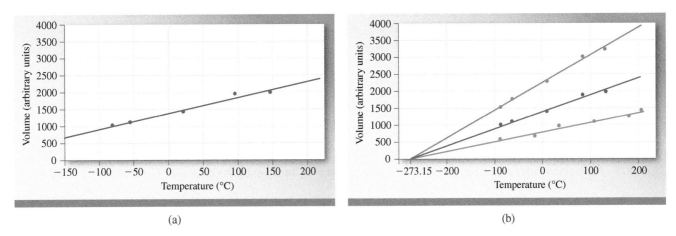

(a)                                                   (b)

**Figure 2** (a) plot showing the relationship between the volume and temperature of an ideal gas (Charles's law). (b) *V* vs. *T* plots at different fixed pressures all converge on a single point at −273.15°C—this point, known as absolute zero, is the temperature where theoretically a gas will shrink to zero volume.

Interestingly, if one collects volume vs. temperature data at different fixed pressures (Figure 2, right), all the lines converge on a common point at −273.15°C. Theoretically at this temperature, a gas will occupy zero volume; this is not observed since individual gas particles have volume and attractive interactions (intermolecular forces) that lead them to condense or form solids at low temperatures. In 1848, the Scottish mathematician/physicist William Thompson (who later became Lord Kelvin) defined −273.15°C as "absolute zero," the lowest attainable temperature, and established the Kelvin temperature scale, where each degree Celsius is equivalent to 1 Kelvin (K).[3] To convert from °C to K, simply add 273.15 to the °C value.

$$-273.15°C = 0 \text{ K}$$
$$0°C \;\;\; = 273.15 \text{ K}$$
$$100°C \;\;\; = 373.15 \text{ K}$$

3. *Amontons's Law:* An important relationship for gases, which often goes unnamed, describes how pressure is related to temperature for a fixed amount of gas under constant volume conditions. This relationship was studied by the French physicist Guillaume Amontons in the late 1600s; Amontons found that the pressure of a gas increases as the temperature is increased (Figure 3):

$$P \, \alpha \, T \qquad\qquad P = k'' \times T \qquad\qquad \textbf{Equation 3}$$

This behavior is observed on long car trips and is well recognized by car racing professionals. As the tires on the car warm up, the pressure in the tires increases. The pressure-temperature relationship was later used by Johann Lambert in 1779[4] to define absolute zero through extrapolation of the data to a point where the pressure is equivalent to zero (this predated the work by Lord Kelvin). Amontons is perhaps best known for his studies of friction, although several inventions are also credited to him, including a pressure-independent air thermometer, the first telegraph, and a type of barometer designed for use on ships.

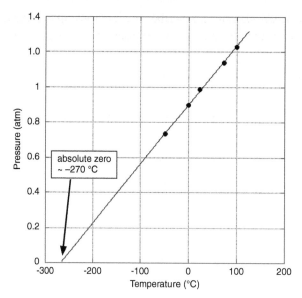

**Figure 3** **Plot showing relationship between pressure and temperature of a gas (Amontons's law).**

4. *Avogadro's Law/Theory/Hypothesis:* The Italian physicist Lorenzo Romano Amedeo Carlo Avogadro, Count of Quaregna and Cerreto, started his professional life in a legal career (following in the path of many in his family), but was eventually drawn away by his interest in math and physics. In 1811, he published a paper[5] in which he put forth the hypothesis that equal volumes of different gases at the same fixed pressure and temperature will all contain equivalent numbers of molecules (i.e., Avogadro's law).

$$V \alpha n \qquad\qquad V = k''' \times n \qquad\qquad \textbf{Equation 4}$$

Under what are known as standard conditions or STP (273.15 K temperature and 1 atmosphere pressure), the volume of 1 mole of any ideal gas is equivalent to 22.414 L (called the standard molar volume); at 25°C, the standard molar volume is 24.465 L mol$^{-1}$ (again at 1 atmosphere pressure).

## THE IDEAL GAS EQUATION

No real gas obeys the gas laws exactly. Most, however, come close under STP conditions, but show deviations as they approach conditions of high $P$ ($> 10$ atm) or low $T$ ($< 200$ K), where particle-particle interactions become more significant. An ideal gas is a theoretical gas that *does* obey all gas laws under all temperature and pressure conditions. For this to be true, the individual particles (1) are assumed to have no volume while still maintaining their mass (i.e., they are "point masses") and (2) do not interact with each other, even at high pressures and low temperatures. The ideal gas is a reasonable approximation for many gases under typical conditions.

By combining the individual gas laws, we can generate a single expression that relates $T$, $P$, $V$, and $n$ for a gas:

$$V \alpha \frac{n \times T}{P} \qquad\qquad V = R \times \frac{n \times T}{P} \qquad\qquad P \times V = n \times R \times T \qquad\qquad \textbf{Equation 5}$$

This expression is known as the *ideal gas law*. The variable $R$ is known as the ideal gas constant and is equal to 0.08206 L atm mol$^{-1}$ K$^{-1}$ (for other applications, $R$ can be converted to 8.314 J mol$^{-1}$ K$^{-1}$).

## MEASURING PRESSURE

What *is* pressure anyway? A gas in a container is exerting pressure on the inner walls of the container, while the air outside the container is doing the same on the outer walls. Pressure is defined as a force per unit area (e.g., per square inch):

$$P = \frac{F}{A}$$

**Equation 6**

The force is created in a gas by the repeated collisions of gas particles against the walls of the container; the more collisions in a given amount of time (per unit area), the higher the pressure; the harder the collisions (caused by the particles moving faster), the higher the pressure. There are many units used to describe pressure, each with significant use:

1 atmosphere* = 101,325 pascals (Pa) [the SI unit for pressure; 1 Pa = 1 Newton m$^{-2}$
= 1 kg m$^{-1}$ s$^{-2}$]

= 1.01325 bar

= 14.7 pounds per square inch (psi)

= 760 mmHg*

= 760 tor*

\* indicates exact number

Pressure is measured using a device called a manometer (for atmospheric pressure applications it is called a barometer). Two simple types of manometers are shown in Figure 4.

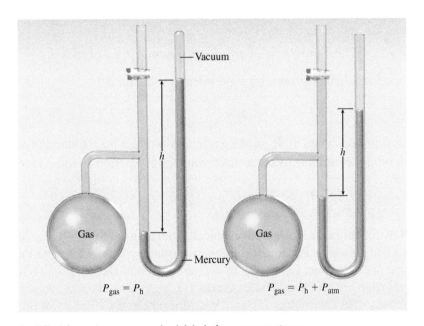

**Figure 4    Closed-ended (left) and open-ended (right) manometers.**

From the relative heights of the two sides of the liquid in the manometer, one can deduce the pressure exerted by the gas(es) in the chamber, recognizing that in an open-ended manometer, the force exerted by the gas in the vessel is against atmospheric pressure.

Close-ended:       $P_{gas} = P_{Hg}$                    **Equation 7**

Open-ended:        $P_{gas} = P_{Hg} + P_{atm}$          **Equation 8**

The pressure exerted by a column of fluid is:

pressure = (density) × (height) × (gravitational constant)     **Equation 9**

where density is expressed in kg m$^{-3}$, height is in meters, and the gravitational constant = 9.80665 m s$^{-2}$.

When liquids other than mercury are used (with corresponding different densities), the height of the liquid column in the manometer/barometer will be different and inversely proportional to the liquid's density. For a gas sample, the column height can be converted from one liquid to another by modifying Equation 9 to give Equation 10.

(height of column)$_{Liq\ A}$ × (density)$_{Liq\ A}$ = (height of column)$_{Liq\ B}$ × (density)$_{Liq\ B}$     **Equation 10**

## THIS LAB

In this lab, there are two goals: (I) build a water manometer and use it to verify Boyle's law (pressure vs. volume) for a sample of air, and (II) determine experimentally the value of the ideal gas constant $R$. For Part II, you will be generating hydrogen gas through a *single displacement reaction* (a redox reaction) between Zn metal and HCl:

$$Zn(s) + 2HCl(aq) \rightarrow H_2(g) + ZnCl_2(aq)$$     **Reaction 1**

[Interesting side note: this reaction was used by Henry Cavendish (English chemist and physicist) in the early 1760s to produce a gas he called "inflammable air," which he determined was a distinct element different from common air;[6] this gas was later renamed "hydrogen" (means "water former") by Antoine Lavoisier, although Cavendish is given credit for its discovery.] From a known mass of Zn and an excess of HCl(aq), the moles of $H_2$ can be determined from a balanced chemical equation. The $H_2$ is generated in a closed container with tubing running to a graduated cylinder filled with water; the $H_2$ bubbles pass through the tubing and through the water into the cylinder, displacing water. The pressure of the $H_2$ can then be determined using Dalton's law and the partial pressure of water at the appropriate temperature:

$$P_{total} = P_{atm} = P_{H_2} + P_{H_2O}$$     **Equation 11**

Using the ideal gas law, one can then determine an experimental value for $R$:

$$R = \frac{P \times V}{n \times T}$$     **Equation 12**

where $P$ is $P_{H_2}$, $V$ is the volume occupied by $H_2$ in the graduated cylinder (i.e., amount of water displaced by $H_2$), $n$ is the number of moles of $H_2$ determined from the stoichiometry of reaction 1, and $T$ is room temperature.

## References

1. Boyle R. (1660). New Experiments Physico-Mechanicall, Touching the Spring of the Air, and Its Effects (Made, for the Most Part, in a New Pneumatical Engine) Written by Way of Letter to the Right Honorable Charles Lord Vicount of Dungarvan, Eldest Son to the Earl of Corke (Hall H, Ed.), Oxford.
2. Gay-Lussac JL. Recherches Sur La Dilatation Des Gaz Et Des Vapeurs. Annales de chimie XLIII. 1802: 137.
3. Thomson W. On an Absolute Thermometric Scale Founded on Carnot's Theory of the Motive Power of Heat, and Calculated from Regnault's Observations. *Math. and Phys. Papers.* 1848; 1: 100–106.
4. Lambert JH. (1779). *Pyrometrie Oder Vom Maaße Des Feuers Und Der Wärme.* Haude und Spener, Berlin.

5. Avogadro A. Essai D'une Maniere De Determiner Les Masses Relatives Des Molecules Elementaires Des Corps, Et Les Proportions Selon Lesquelles Elles Entrent Dans Ces Combinaisons. *Journal de Physique.* 1810; 73: 58–76.

6. Cavendish H. Three Papers Containing Experiments on Factitious Air. *Philosophical Transactions.* 1766; 56: 141–184.

7. *Handbook of Chemistry and Physics (CRC).* 73rd edition. 1992. Lide, E. R., ed. Chemical Rubber Publishing Company. Boca Raton, FL.

## *Procedure*

### I. Boyle's Law Experiment—Dependence of Volume on Pressure

1. Build a water manometer to measure pressure (Figure 5):
    a. Use a piece of Tygon tubing (at least 48 in long)—the inner diameter of the tubing should be measured or obtained from the packaging; the tubing that you select should attach firmly to the end of a 50-mL plastic syringe.
    b. Arrange the tubing in a "U" shape and attach in a vertical position. One side of the "U" should be slightly longer (this will be the side that you attach the syringe to). The distance between the two vertical arms should be close enough to allow both to overlap a sheet of graph paper placed behind the tubing; options to construct this include (1) using clamps to hold the tubing in a "U" shape on a ring stand, (2) attaching it to a vertical piece of plywood using clamps or tape or something similar, and (3) taping the tubing to a wall or the sash of a hood in the lab.
    c. Tape a sheet of graph paper behind, being sure to arrange with the lines as perpendicular to the floor as possible.
    d. Add water through one arm of the tubing using a water bottle, filling the "U" approximately half of the distance up each arm.

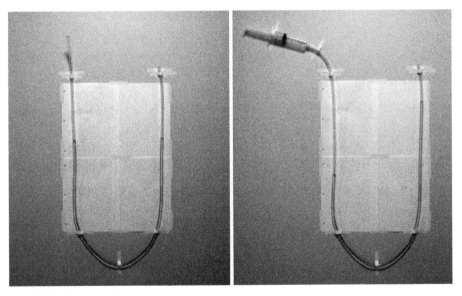

**Figure 5  Water manometer.**
Left image shows U-shaped tube filled with water colored with food coloring. Right image has a 50-mL syringe attached and the plunger depressed to reduce volume.

2. Draw back the plunger of the syringe, filling it with exactly 50 mL of air. Then carefully attach the syringe to the long arm of the manometer.

3. On the side with the syringe, measure the distance from the top of the water to the end of the syringe.

4. On the graph paper, mark the position of the top of the water on both arms—label these with a "50".
5. Slowly push the plunger in to a new volume position; allow the manometer to respond, then mark the new position of the top of the water on both arms with the syringe volume (e.g., "45").
6. Repeat step 5 in small increments until either (1) the syringe is empty (i.e., reads "0 mL") or (2) the water in the shorter arm of the manometer approaches the top of the tubing. You should have at least 12 measurements.
7. Use a ruler to measure the vertical distance between each pair of marks (e.g., the distance between the two "50" markings)—record these values (in millimeters).

## II. Ideal Gas Law Experiment—Experimental Determination of Ideal Gas Constant *R*

8. Attach one end of a piece of rubber hose or Tygon tubing to a short piece of glass tubing that has been inserted through the hole of a rubber stopper; the other end of the rubber hose should be placed on the long end of a piece of glass tubing that has a short "U"-shaped hook at the other end (Figure 6).

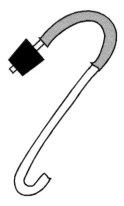

**Figure 6   Rubber stopper / hose / "U"-shaped tubing assembly.**

9. Set up the H$_2$ gas collector:
   - Immerse and completely fill a 600-mL beaker and a 100-mL graduated cylinder in a large container of water (Figure 7A; the water container could be a filled bucket, or even a sink plugged and filled with water).
   - While it is still immersed, invert the graduated cylinder and insert it into the beaker (Figure 7B).

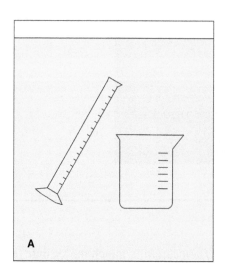
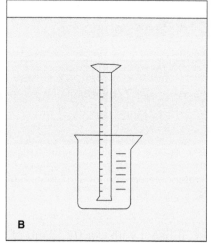
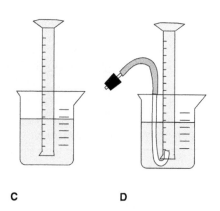

**Figure 7   H$_2$ gas collector setup—see text for details.**

- Lift both from the water, being careful to keep the graduated cylinder inserted in the beaker and filled with water (*prevent air from getting into the cylinder*); pour some water from the beaker, keeping 350 to 400 mL in the beaker (Figure 7C).
- Place the beaker/cylinder on the counter next to the ring stand.
- CAREFULLY insert the U-shaped end of the glass tubing into the water in the beaker and insert under the graduated cylinder—you may lift the cylinder just enough to get the end of the tubing into the cylinder's mouth; again, *do not allow air into the cylinder* (Figure 7D).

10. Set up the $H_2$ gas generator:
    - Attach test tube to the ring stand at ~45° angle using a clamp.
    - Add 8 mL of 6.0 *M* HCl to the test tube.
11. Obtain and determine the mass of a small piece of Zn metal (mass should be ~0.200 to 0.225 g).
12. Cut a piece of filter paper into four quarters; wrap the piece of Zn using one piece of filter paper; wet with water to hold the filter paper closed, and then press to remove excess water.
13. With care, insert wrapped piece of Zn into test tube approximately 1 in so that it adheres to the side of the tube wall.
14. Carefully and securely insert the rubber stopper into the test tube (Figure 8).

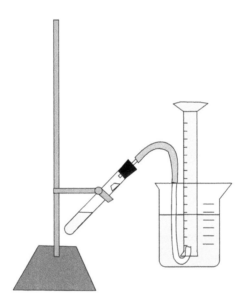

Figure 8   **Final setup, with Zn and HCl(*aq*) in test tube.**

15. To initiate the reaction, gently tap the clamp holding the test tube with a pen (do not hit the glass test tube)—this should dislodge the wrapped Zn and cause it to slide into the HCl(*aq*). $H_2$ gas should begin bubbling into the beaker and filling the graduated cylinder; this process will continue for as long as 20 min.
16. After the reaction is complete (i.e., no more gas is being generated), carefully remove the "U"-shaped tubing from the beaker, taking care to not allow air to get into the cylinder.
17. Again submerge the beaker/graduated cylinder setup completely into the large container of water; remove the beaker from the bottom of the cylinder (again, do not let air into the cylinder). Now, carefully adjust the height of the cylinder so that the level of the water outside the cylinder exactly matches the water level inside the cylinder—this ensures that the total pressure inside the cylinder matches the pressure of the room (Figure 9).
18. Record the volume of the gas in the graduated cylinder.

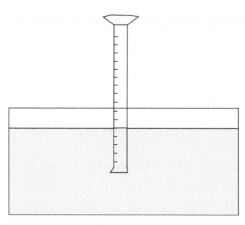

**Figure 9 Position of graduated cylinder is adjusted until water level inside the cylinder matches the water level outside the cylinder.**

19. You may now remove the cylinder from the water, keeping it inverted; take it to a hood and turn upright to release the $H_2$ gas harmlessly into the hood.
20. Record the atmospheric pressure and the temperature of the water in the large water container—these will be used in our ideal gas law calculation.
21. The HCl($aq$) can be poured into the appropriate waste container.

## Data Collection

### I. Boyle's Law Experiment

1. Inner diameter of the Tygon tubing: _____ cm
2. Distance from the top of the water to the bottom of the syringe: _____ cm
3. Vertical distance between each pair of marks:

| Volume in Syringe (mL) | Vertical Distance Between Marks (mm) |
|---|---|
|  |  |
|  |  |
|  |  |
|  |  |
|  |  |
|  |  |
|  |  |
|  |  |
|  |  |
|  |  |
|  |  |
|  |  |
|  |  |
|  |  |

## II. Ideal Gas Law Experiment—Experimental Determination of Ideal Gas Constant $R$

**4.** Mass of Zn metal: _____ g

**5.** Volume of $H_2$ gas: _____ L

**6.** Atmospheric pressure: _____ atm

**7.** Water temperature: _____ °C

## Analysis

### I. Boyle's Law Experiment

**8.** Volume of air in syringe side of manometer $= (\text{length}) \times \pi \times (\text{inner radius})^2$

$$= (\#2) \times \pi \times [0.5 \times (\#1)]^2$$

$$= \underline{\hspace{2cm}} \text{ cm}^3$$

**9.** Total volume at each step $= (\text{volume in syringe}) + (\text{volume in long arm of manometer})$

$$= (\text{volume in syringe}) + (\#8)$$

| Step (volume in syringe, mL) | Total Volume (cm³ or mL) |
|---|---|
| | |
| | |
| | |
| | |
| | |
| | |
| | |
| | |
| | |
| | |
| | |
| | |
| | |
| | |
| | |
| | |

**10.** Convert vertical distance between marks (i.e., height difference of water on two sides of manometer) to pressure:

$$\text{Pressure (atm)} = \frac{(\text{height of column of water in mm})}{(13.5951) \times (760)} \qquad \textbf{Equation 11}$$

$$= \frac{(\#3 \text{ value})}{(13.5951) \times (760)}$$

| Step (volume in syringe, mL) | Pressure (atm) | 1 / Pressure |
|---|---|---|
| | | |
| | | |
| | | |
| | | |
| | | |
| | | |
| | | |
| | | |
| | | |
| | | |
| | | |
| | | |
| | | |
| | | |
| | | |

11. Plot [Total volume (L) in #9] vs. [1 / pressure (atm) in #10]:

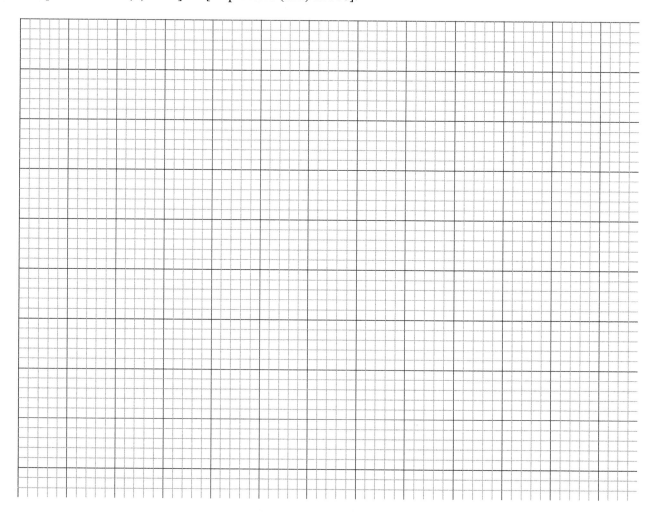

12. Using Excel, a graphing calculator, or the least-squares method you utilized in Laboratory 4, fit your data above to a straight line of the form:

$$\text{volume} = (\text{slope}) \times \left(\frac{1}{P}\right) + (\text{y-intercept})$$

  a. Slope of your line (including units): _____

  b. *Y*-intercept of your line (including units): _____

## II. Ideal Gas Law Experiment—Experimental Determination of Ideal Gas Constant *R*

13. Moles of Zn metal: _____ mol

14. Moles of $H_2$ produced: _____ mol

15. Partial pressure of water: _____ atm     (use Table 1 below and interpolation)

| Table 1    Vapor Pressure of Water at Different Temperatures.[7] | | | |
|---|---|---|---|
| Temperature (°C) | Water Vapor Pressure (mm Hg) | Temperature (°C) | Water Vapor Pressure (mm Hg) |
| 0 | 4.5840 | 28 | 28.376 |
| 5 | 6.5449 | 29 | 30.071 |
| 10 | 9.2123 | 30 | 31.855 |
| 11 | 9.8483 | 31 | 33.730 |
| 12 | 10.522 | 32 | 35.700 |
| 13 | 11.237 | 33 | 37.769 |
| 14 | 11.993 | 34 | 39.942 |
| 15 | 12.795 | 35 | 42.221 |
| 16 | 13.642 | 36 | 44.613 |
| 17 | 14.539 | 37 | 47.121 |
| 18 | 15.487 | 38 | 49.750 |
| 19 | 16.489 | 39 | 52.506 |
| 20 | 17.546 | 40 | 55.391 |
| 21 | 18.663 | 45 | 71.968 |
| 22 | 19.841 | 50 | 92.648 |
| 23 | 21.085 | 60 | 149.61 |
| 24 | 22.395 | 70 | 234.03 |
| 25 | 23.776 | 80 | 355.63 |
| 26 | 25.231 | 90 | 526.41 |
| 27 | 26.763 | 100 | 760.00 |

**16.** Pressure of $H_2$ gas $= P_{atm} - P_{H_2O}$

$$= (\#6) - (\#15) = \underline{\hspace{3cm}} \text{ atm}$$

**17.** Experimental $R = \dfrac{P \times V}{n \times T} = \dfrac{(\#16) - (\#5)}{(\#13) \times (\#7 \; in \; K)}$

$$\underline{\hspace{3cm}} \text{ mol}$$

**18.** % error (experimental $R$ vs. literature value for $R$): $\underline{\hspace{3cm}}$

## Reflection Questions

1.  (a) In your plot of $V$ vs. $1/P$, what is the numerical value for the slope?

    (b) Based on the ideal gas law, what does the slope of this best-fit line represent?

    (c) What do you expect the $y$-intercept of the best-fit line to equal?

    (d) If you assume the temperature in the room was 20.0°C when you conducted the Boyle's law experiment, how many total moles of gas were in the trapped volume (i.e., within the syringe plus within the tubing between the water and the syringe)?

2.  Using ideas from the kinetic molecular theory, explain why pressure increases as volume decreases.

3. Let's imagine that you used ethanol (density $= 0.789$ g cm$^{-3}$) instead of water for your manometer. Determine the pressure (in millimeters ethanol) that your manometer would record for the maximum pressure that you observed in Part I.

4. Imagine that the piece of Zn metal that you obtained for part II was coated with a thin oxide layer, ZnO. ZnO will also react with HCl:

$$ZnO(s) + HCl(aq) \rightarrow H_2O(l) + ZnCl_2(aq) \qquad \text{(unbalanced)}$$

(a) Write the balanced molecular, ionic and net ionic equations for this reaction:

(b) What would happen to the moles of $H_2(g)$ you collect if the ZnO layer was present? How about the value of $R$ that you would calculate?

5. Sodium reacts with water to produce $H_2(g)$ and sodium hydroxide in a highly exothermic reaction:

$$2Na(s) + 2H_2O(l) \rightarrow 2NaOH(aq) + H_2(g)$$

The heat from the reaction is sufficient to ignite the hydrogen gas. Assume this reaction was carried out under conditions that prevent $H_2$ ignition. Using the same mass of Na$(s)$ as was used for Zn$(s)$ in Part II of your experiment, the same water temperature, and the same atmospheric pressure, determine the volume of $H_2(g)$ that you would generate, showing all of your calculations:

6. What will be the effect of each of the following experimental errors on the value of $R$ that you calculate in Part II?
   • After collecting the hydrogen gas, you do not adjust the height of the graduated cylinder appropriately—the height of the water inside the cylinder is lower than that of the water outside the cylinder when you record the volume:

- You forget to correct for the pressure of water vapor in the graduated cylinder:

- You do not completely react all the Zn metal:

## *Connection*

Based on your experience in this lab, draw a connection to something in your everyday life or the world around you (something not mentioned in the background section):

# 14

# A Capstone Experience: Toward the Creation of an Automobile Airbag

## Objectives

- Integrate general chemistry I concepts to create a model for an automobile airbag.
- Determine a method and then carry out appropriate procedures to inflate a bag with a gas, given a specified set of materials. NOTE: *You are not to share details of this lab with those outside of your lab section so that all students can have the same experience as you.*

## Materials Needed

- Resealable plastic bags (e.g., Ziploc bags)
- Baking soda (sodium bicarbonate, $NaHCO_3$)
- 6 $M$ acetic acid ($HC_2H_3O_2$)
- 100-mL graduated cylinder
- Water
- Plastic spoon or metal spatula
- 1-L beaker
- Calculator
- Ruler

## Safety Precautions

- Safety goggles/glasses must be worn at all times.
- 6 $M$ acetic acid is toxic by ingestion and inhalation and potentially corrosive to skin and eyes; avoid contact with body tissues.

## Background

After surviving a car accident with his wife and young daughter, John Hetrick, a retired engineer, developed and ultimately patented the first automobile airbag.[1-3] In the latter part of that same decade, the automotive industry began to take notice and developed several early airbag designs, though airbags did not make it into commercial vehicles until the early 1970s and weren't commonly included as standard equipment in all automobile models until the 1990s. The idea was quite simple: provide an inflatable cushion to buffer the effects of an automobile crash on the human body. The challenge was not, hence, the lengthy time between the original Hetrick patent and general commercial use. In short, one needed to produce a cushion in fractions of a second with no possibility of "non-accident" deployment while using substances that are not harmful to the people being protected. The answer to this challenge involved chemistry.

## HOW DOES AN AIRBAG WORK?

Sensors in a car detect rapid deceleration (as resulting from a head-on car crash) and send an electrical signal that initiates a chemical reaction, which, in turn, produces the gas $N_2$. The gas inflates the airbag at approximately 200 miles per hour—enough force to potentially injure the passenger! After full inflation, vents in the airbag allow for the pressurized nitrogen to escape the airbag, which is a key to its proper use. The airbag is designed such that the passenger will contact it as it is deflating (not inflating!), thereby providing a cushion. The entire time of this process is measured in milliseconds.

## THE CHEMISTRY

Many airbags, in particular earlier models, function using sodium azide, $NaN_3$, as the propellant. In this case, the overall balanced chemical reaction that causes the inflation of the automobile airbag is given below:

$$10NaN_3(s) + 2KNO_3(s) \rightarrow 5Na_2O(s) + K_2O(s) + 16N_2(g) \qquad \textbf{Reaction 1}$$

This reaction occurs in two steps, shown in reactions 2 and 3 below.

$$2NaN_3(s) \rightarrow 2Na(s) + 3N_2(g) \qquad \textbf{Reaction 2}$$

$$10Na(s) + 2KNO_3(s) \rightarrow K_2O(s) + 5Na_2O(s) + N_2(g) \qquad \textbf{Reaction 3}$$

The first reaction involves the thermal decomposition of sodium azide into metallic sodium and nitrogen gas (used to inflate the airbag). The second reaction consumes the highly reactive sodium metal, converting it into the corresponding oxide and producing more nitrogen gas. A third reaction uses silicon dioxide to convert the oxides of potassium and sodium, the products of Reaction 3, into an inert, nontoxic material, glass. The final chemical products are nitrogen gas and glass, a very stable, nonflammable substance.

$$K_2O(s) + Na_2O(s) + SiO_2 \rightarrow glass \qquad \textbf{Reaction 4}$$

Importantly, engineers need to perform stoichiometric calculations to ensure that the appropriate amount of starting materials are loaded into the airbag mechanism to produce exactly the right amount of nitrogen to fully inflate the bag. This is a critical concept for you to understand in developing procedures to achieve the goal of this lab.

## References

1. John Hetrick, John. Pat. No. 2,649,311. Safety Cushion Assembly for Automotive Vehicles (filed Aug. 18, 1953).
2. Madlung, A. The Chemistry Behind the Air Bag: High Tech in First-Year Chemistry *J. Chem. Ed.*, 1996; 73(4): 347–348.
3. Scoltock, J. John Hetrick *Automotive Engineer,* 2011; 36(1): 7.

## Procedure/Data Collection/Analysis

Before you can begin this laboratory, *you* must develop a procedure to achieve the desired goal, which is to inflate a Ziploc bag with a gas such that the bag is filled to its *exact* maximum capacity. You may assume the lab temperature and pressure to be 25.0°C and 1.00 atm, respectively.

1. Given the listed materials and standard resources such as a balance, periodic table, etc., devise a step-by-step procedure that clearly describes how you will accomplish the experimental goal. (*Note:* A good set of procedures means that one of your classmates could use them to perform this experiment and achieve the identical result as you.)

   Factors that you need to consider are as follows:
   - A chemical reaction that produces a gas is required.
   - A balanced equation describing the reaction is required.
   - A process to determine the appropriate amounts of starting materials is necessary.
     (Hint: Consider reaction stoichiometry and the ideal gas law.)

**2.** List any chemical reaction(s) that must take place for you to accomplish the experimental goal.

**3.** Calculations: neatly include any calculations that you need to accomplish the experimental goal.

**4.** Perform the experiment and note the observations here along with any modifications to your procedures. For example, is your reaction endothermic or exothermic (how do you know)? Did a gas evolve? How quickly?

**5.** Did the reaction go to completion? How do you know?

**6.** Describe the appearance of your airbag at the end of the experiment. Did you meet your goal? (*Have the instructor look at the inflated airbag and initial your description.*)

**7.** Describe any errors associated with your experiment.

8. Could you actually use this reaction to inflate an automobile airbag? Discuss briefly any pros and cons associated with its potential use.

9. Considering that a typical driver-side airbag contains 50.0 g of sodium azide, what volume of nitrogen gas would be produced upon impact assuming sodium azide is the limiting reactant? You may assume a temperature of 25 °C and an atmospheric pressure of 760 mm Hg. (*Note:* Sodium azide is quite toxic, furthering the argument for its use as the limiting reactant.)

## Connection

Based on your experience in this lab, draw a connection to something in your everyday life or the world around you (something not mentioned in the background section). You may wish to consider an application for the "airbag chemistry" outside of automobiles.

# 15

# On the Nature of Solutions: Polarity, Energy, and Properties

## Objectives

- Review molecular shape and polarity.
- Understand the principles and use of thin layer chromatography.
- Demonstrate solution properties involving a nonvolatile solute.
- Determine the heat of solution.

## Materials Needed

- Plastic weigh boats (4)
- 2-mL volumetric pipette
- 10-mL graduated cylinder
- Small Erlenmeyer flask (25-mL)
- Methanol
- Ethylene glycol
- Balance
- Dichloromethane
- Benzoic acid
- Ethyl benzoate
- Benzyl alcohol
- Benzylamine
- Toluene
- Small vials (5)
- Silica TLC plates with fluorescent indicator
- TLC chamber (e.g., small jar with lid, or beaker with watch glass)
- TLC spotters
- Handheld UV lamp
- 2 Styrofoam coffee cups and 1 lid
- Thermometer or temperature probe
- Magnetic stir bar
- Magnetic stirring plate
- Hot plate
- 100-mL graduated cylinder
- 600-mL beaker
- 50-mL Erlenmeyer flask
- 4 to 4.5 g Ammonium nitrate ($NH_4NO_3$) or calcium chloride ($CaCl_2$)
- Deionized water

## Safety Precautions

- Gloves and safety goggles/glasses should be worn during this lab.
- Use of a laboratory fume hood is recommended for the TLC component of this lab to avoid inhalation of volatile organic substances.
- The organic substances in this lab are flammable. Keep away from open flames.
- Avoid directing the UV lamp toward anyone's eyes (including your own!).
- Dispose of waste solutions in designated waste containers as indicated by your instructor.

## Background

Solutions . . . ubiquitous in nature and necessary for most chemical reactions! Why are most forms of matter in and around us mixtures? Why are mixtures homogeneous (solutions) in some cases, but heterogeneous in others. Ultimately, as with all properties and functions of matter, the answer lies in the fundamental structures of the components that comprise a particular mixture. From the combination of elements with their respective properties to the types and numbers of bonds in a structure, a staggering assortment of molecules and ionic compounds exist. How these compounds interact is the foundation to understanding solution formation.

Before we explore this notion further, recognize that the very idea of mixtures being more common than separated pure substances is a natural phenomenon that relates to a mixed state being a more disordered state. Nature prefers disorder. Consider how much energy you exert to create order in your daily life: cleaning your car, room, or home; organizing notes and notebooks; arranging contents in purses and wallets; and so on. If you do not exert this effort, the natural process of disorder will take effect without you having to do any work (with obvious results). If a deck of playing cards is dropped, what is the likelihood that each card will land in a single stack with perfect numerical and suit sequence? In contrast, how many other arrangements, each representing a disordered state, are possible? There are simply more ways for mixtures to exist, or more disordered arrangements, than there are for pure substances, which makes mixtures more likely to occur and therefore more dominant in our environment.

However, not all substances mix. Cooking oil and water certainly don't form a solution. While recognizing that nature prefers a mixed state, there are clearly situations in which other factors, factors that are always present, become dominant. These other factors relate directly to chemical structure. A guiding principle for solution formation is "like dissolves like," meaning polar solutes interact more strongly with polar solvents and nonpolar solutes interact preferably with nonpolar solvents. Cooking oils, comprised of molecules that contain primarily nonpolar C—C and C—H bonds, are nonpolar substances. Water, of course, is a bent-shaped molecule with very polar O—H bonds due to the electronegativity difference between O and H atoms; the result is a polar molecule containing fixed and significant partial positive and negative sides. Thus, it is the different polarities of oil and water that prevent their mixing. Each has more favorable interactions between their own respective molecules than those with a poorly matched counterpart in terms of polarity. You may recall that molecular polarity was determined by a process that begins with a successful Lewis structure, followed by the application of VSEPR theory and, coupled with an understanding of electronegativity, analysis of the resultant three-dimensional arrangement of the polar bonds. This leads to the assignment of intermolecular forces, which, in turn, govern the properties of molecular substances (surface tension, vapor pressure, viscosity, boiling points, etc.) and the formation of solutions (Figure 1).

Solution formation is accompanied by a change in energy (or enthalpy under conditions of constant pressure, such as atmospheric pressure). The factors that contribute to the energy change are structure-based. Considering the final state of a solution, a homogeneous mixture in which both solvent and solute particles are dispersed and interacting, the energetics of solution formation can be broken down into three contributing pieces. First, the solvent must be expanded to create space for the solute particles. This endothermic process requires separating solvent particles from each other. Similarly, because the solute is distributed throughout the solvent in a solution, the solute particles have to be separated as the solution forms, another energy-requiring

**Figure 1** A six-spotted fishing spider at rest on the water's surface in Chillicothe, Missouri. The strong hydrogen bonding present in water creates sufficient surface tension to allow the spider to rest, walk, and hunt on water.

or endothermic process. The third piece focuses on the energy that accompanies the combination of the solute and solvent particles to form the solution. It is this step that gives us the simple "like dissolves like" description for solution formation. In other words, a polar solute and polar solvent will combine readily through favorable intermolecular attractions and release energy into the environment in an exothermic process. The same can be said for the favorable interactions of two nonpolar components. If either of the first two endothermic steps are energetically prohibitive in comparison to the exothermic third step, then the solution will likely not form (e.g., oil and water). Or, put differently, if the third step has an unfavorable match in polarity between the two components, then the energy it provides will be minimal and solution formation is unlikely. However, if the third step provides sufficient energy to significantly counter the endothermic requirements of the first two steps through "like" interactions (polar-polar or nonpolar-nonpolar), then solution formation is likely. The sum of the three steps to solution formation is measurable and termed the "heat of solution" or $\Delta H_{soln}$. Heats of solution can be endothermic (solution formation in this case is driven largely by a preference for disorder) or exothermic. In fact, commercially sold hot and cold packs function by the formation of solutions with exo- and endothermic heats of solution, respectively (Figure 2).

In summary, the formation of solutions is governed by energetics as linked directly to chemical structure (polarity) and a natural tendency toward disorder. In this lab, you will explore the relationship between chemical structure and polarity using a technique called thin layer chromatography or TLC. You will then investigate the impact that a nonvolatile solute has on the evaporation rate of a volatile solvent, a starting point for understanding that solution properties are different than those of pure substances. Finally, using calorimetry, you will measure the heat of solution for an ionic compound dissolved in water.

## THIS LAB

This lab uses three separate tasks, each described below, to highlight fundamental aspects of solutions: polarity, energetics, and a representative solution property (vapor pressure lowering).

### I. Thin Layer Chromatography

Thin layer chromatography (TLC) is a technique that is used to separate components in a mixture or to identify a pure component based on molecular polarity and, hence, chemical structure. TLC involves a solid phase and a mobile phase, silica ($SiO_2$) and methanol/dichloromethane, respectively, in today's experiment. In this technique, a small sample is placed or "spotted" on the bottom of a silica-coated TLC plate. The plate is then placed in a

**Figure 2** Ammonium nitrate is a type of salt found in instant cold packs because it has an endothermic heat of solution (i.e., absorbs heat on mixing with water).

TLC chamber, a jar or beaker, containing a shallow level of liquid eluent (solvent or the mobile phase), such that the spot is above the surface of the eluent (Figure 3). A cover is then placed on top of the chamber. As the eluent slowly migrates up the plate, by capillary action, the spotted sample desorbs from the silica surface (the solid phase) and moves up the plate. The degree of interaction between the sample and the solid surface is a key factor in the experiment. The silica surface is polar. As such, more polar analytes (substances under analysis) interact with the solid surface to a greater degree and move more slowly up the plate. Nonpolar substances migrate faster, thus, resulting in separation based on molecular polarity. To interpret results, the plate must be "developed." In this lab, a handheld ultraviolet lamp shined on the plate (not in your or another's eyes!) will allow you to see the final

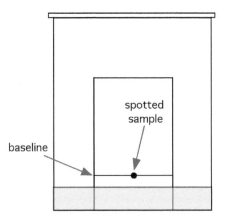

**Figure 3** The starting point for a TLC experiment. Note the level of the eluent is below that of the spotted sample.

position of the spots. An increase in the polarity of the eluent or mobile phase will cause all samples, regardless of polarity, to migrate faster up the plate. A decrease in the polarity of the eluent will have the opposite effect. Thus, the nature of the eluent gives great control to the experimentalist seeking to identify a pure substance or the components in a mixture.

The five analytes in today's lab are organic molecules that all contain a hydrophobic phenyl group, the six-membered ring shown in each structure (Figure 4). Please note that not all elements are shown in these structures. This shorthand notation for describing organic compounds does not include H atoms attached to C atoms, nor does it include the actual C atom labels. You should also be reminded that C—C and C—H bonds are viewed as effectively nonpolar bonds. Thus, each analyte differs in the remainder of the structure. By inspection of the polarity of the unique structural elements in each structure, you should be able to predict and then confirm through experimentation their respective migrations on a silica TLC plate.

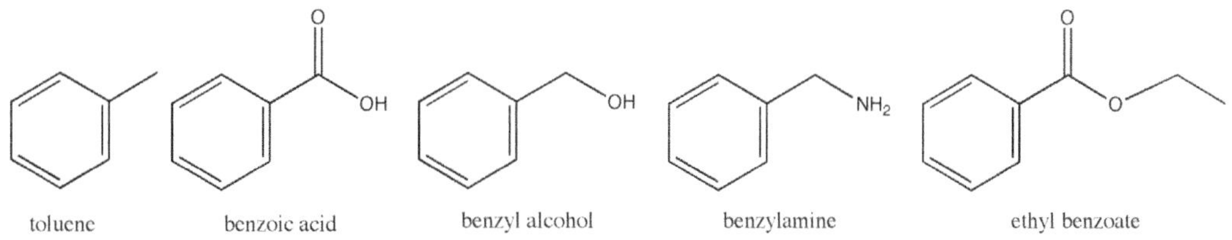

**Figure 4**   The structures of the analytes to be studied by thin layer chromatography, TLC.

## II. The Volatility of a Solution Containing a Nonvolatile Solute

Solution properties are determined by the chemical composition of the solution components and the concentration of the solution (molarity, mole fraction, molality, mass percent). In this part of the lab, you will prepare three solutions with varied amounts of ethylene glycol and methanol and measure the amount of evaporation that takes place in the three solutions, in addition to a sample of pure methanol over a period of 15 min. Ethylene glycol and methanol contain *alcohol* groups (i.e., O—H bonds) that allow for hydrogen bonding, strong intermolecular forces. Not surprisingly, because of their similar structures (Figure 5) and the "like dissolves like" theme for solution formation, they are miscible (soluble) in infinite proportions. At room temperature, ethylene glycol has negligible vapor pressure. Thus, during this experiment, it is to be viewed as a nonvolatile solute, which means a pure sample of ethylene glycol will not evaporate to an appreciable extent over the measurement time frame. Methanol, however, has a much higher vapor pressure and will be the evaporating substance that allows for data to be collected. Prior to performing this experiment, you should have an expectation for the results based on the percentage of the nonvolatile ethylene glycol present in a particular sample.

**Figure 5**   The structures of methanol (left) and ethylene glycol (right).

## III. Determination of the Heat of Solution

Calorimetry is a technique for the measurement of enthalpy changes. In Lab 12, you were introduced to the principles, procedures, and calculations of calorimetry. In this lab, you will be using calorimetry as a tool to determine the heat of solution for either $NH_4NO_3(aq)$ or $CaCl_2(aq)$.

## References

1. Baker TH, Fisher GT, Roth JA. "Vapor Liquid Equilibrium and Refractive Indices of the Methanol-Ethylene Glycol System." *Journal of Chemical and Engineering Data* 1964; 9(1): 11–12.

## Procedure

### I. Thin Layer Chromatography

1. Using the five small vials, prepare five solutions by dissolving a small amount of each substance listed below (a few grains if solid, a drop if liquid) in 10 drops of dichloromethane. Label each vial.

   Vial 1: benzoic acid

   Vial 2: ethyl benzoate

   Vial 3: benzyl alcohol

   Vial 4: toluene

   Vial 5: benzylamine

2. Obtain or cut two TLC plates (plastic-backed silica plates with fluorescent indicators).
3. Place five small marks using a pencil along a line 0.5 to 1 cm from the bottom of each plate. You do not need to draw a line, but you need to have the five pencil marks equally distant from the bottom of the plate. For best results, do not place a pencil mark close to either edge of the plate.
4. Using a glass capillary TLC spotter, spot each of the five samples on a separate pencil mark on each TLC plate. Keep the order of the vials (1–5, as described above) consistent with the "left-to-right" delivery of five spots to avoid confusion as to which substance corresponds to which spot or, more importantly, be sure that you record which substance is on which pencil mark in the Data Collection section below. Your spotting technique should involve the targeted delivery of a small amount of solution directly on each pencil mark. Avoid large spots that mix with neighboring spots.
5. Add enough of a 2% methanol in dichloromethane solution to just cover the bottom of the TLC chamber.
6. Place the TLC plate in the TLC chamber such that the solution (eluent) level remains below the marked spots on the bottom of the TLC plate. If the initial spots (pencil marks) are below the eluent level, then you must begin this experiment again. The top of the TLC plate should rest against the inside wall of the TLC chamber. Place the lid or cover on top of the TLC chamber.
7. Allow the eluent to migrate to approximately 0.5 cm from the top of the TLC plate and then remove the plate from the chamber and mark the solvent or eluent front with a pencil line.
8. Being careful not to shine the UV light in your or other's eyes, look at your TLC plate under UV light. Using a pencil, circle any observed "spots" on the TLC plate.
9. Repeat the previous steps using pure dichloromethane as the eluent.
10. Neatly and accurately reproduce (sketch) your two developed TLC plates on the provided diagrams in the Data Collection section below.

### II. The Volatility of a Solution Containing a Nonvolatile Solute

In this section, you will be working with four liquids, one pure substance and three solutions. Each liquid is to be studied as described below over a 15-min period. For good data, it is critical that the start time and end time for each liquid be a consistent 15-min interval. If you lose track of time, you must start that sample measurement again.

11. Obtain four plastic weigh boats.

12. Using a 2-mL volumetric pipette, transfer 2.00 mL of methanol into one of the plastic weigh boats and record the mass of methanol. You may weigh by difference (record the mass of weigh boat, add methanol sample, record mass of sample plus weigh boat and take the difference as the mass of methanol) or tare the balance with the weigh boat on it and then add the methanol sample.

13. To allow others to use the balance, cover your methanol sample promptly (after recording the mass) with either another weigh boat or a watch glass and carefully move it to your work space. Do not place in a laboratory hood or in a busy laboratory area with considerable or fluctuating air current. Be careful not to splash the methanol sample such that it either spills or contacts the temporary cover. Remove the cover and record the initial time.

14. Because you are using a single 2-mL volumetric pipette for transferring four different liquids, it is imperative that the volumetric pipette is rinsed with methanol, dried in an oven, and allowed to cool prior to each use. Alternatively, it can be rinsed thoroughly with a small amount of the sample for which it is to be used prior to the collection of a 2-mL sample for analysis.

15. After 15 min (record the time), cover your sample and carefully (don't splash or lose sample by contact with the lid) move the sample back to the balance. Remove the cover and record the final mass of your sample. Calculate the percent of the original sample mass that was lost.

16. Using a 10-mL graduated cylinder, prepare a solution containing 25% methanol and 75% ethylene glycol by volume. Add the 2.5 mL of methanol first, followed by 7.5 mL of ethylene glycol. Adding the denser glycol second will help with complete mixing of the solution.

17. Pour the solution into a small Erlenmeyer flask (25-mL) and immediately repeat steps 12 through 14, making sure to place this sample in approximately the same laboratory space as the prior sample.

18. Repeat steps 15 and 16 for a solution containing 50% methanol and 50% ethylene glycol by volume.

19. Repeat steps 15 and 16 for a solution containing 75% methanol and 25% ethylene glycol by volume.

20. Make a plot of mole fraction of methanol (x-axis) vs. the percent of original sample mass remaining (y-axis). Include a fifth point on your plot for pure ethylene glycol (mole fraction of methanol = 0, percent of original sample mass remaining = 100), a nonvolatile solute at room temperature.

III. **Determination of the Heat of Solution**

A. **Determination of the Heat Capacity (Calibration) of a Coffee-Cup Calorimeter**

*Note: The following procedures for constructing and calibrating a coffee-cup calorimeter are reproduced from Lab 12.*

21. Set up a hot water bath by putting ~500 mL of water into a 600-mL beaker; place the beaker on a hot plate and set on high. Pay attention as additional water may need to be added if the water level gets too low. [*Note:* If stirrer/hot plates are in limited supply in the lab, multiple students/teams might consider sharing a larger hot water bath.]

22. Take two coffee cups, nest one inside the other, and determine their combined mass.

23. Add ~70 mL of room temperature deionized water ("cool water") to the nested coffee cups and determine the total mass.

24. Construct the coffee-cup calorimeter (Figure 6):
   • Take two nested Styrofoam coffee cups.
   • Add a magnetic stir bar.
   • Add a single polystyrene lid with 1 small hole (for the thermometer).
   • Place calorimeter on a magnetic stirring plate.

25. Tare a 50-mL Erlenmeyer flask; add ~30 mL of room temperature deionized water to the flask and record the mass of the water (this will be the "hot water").

26. Suspend the Erlenmeyer flask in the water bath, making sure the water in the flask is below the water level in the bath; allow the flask to sit in the water bath for ~10 to 15 min.

27. Use your thermometer to record the temperature of the water in the water bath beaker—this will be the temperature of the hot water.

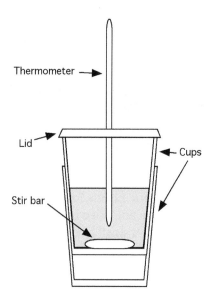

**Figure 6**  Diagram showing assembly of a coffee-cup calorimeter.

28. Before adding the hot water to the calorimeter, insert your thermometer through the hole in the calorimeter lid and use a clamp to secure it—be sure that the bulb of the thermometer is positioned approximately in the middle of the water (i.e., not touching the bottom or sides of the cup); record the temperature (this will be the temperature of the cool water and should be recorded both in line #7 in the Data Collection section, as well as for time = 0 temperature in line #8).

29. After you are sure the thermometer is high enough to not contact the stir bar, turn on the magnetic stirrer; be sure that there is no splashing.

30. Now, use the clamp to remove the Erlenmeyer flask from the bath; lift the calorimeter lid and carefully (but quickly) pour the hot water into the calorimeter—this is time = 0 s. Replace the lid and be sure that the thermometer bulb is in the water.

31. Record the time and temperature every 15 to 30 s for a total period of 8 min. The time between recordings can be increased as the temperature changes become small.

32. Once done, empty the calorimeter and dry completely being careful not to crack the cups or lid.

**B.** **Determination of the Heat of Solution**

33. Add ~100 mL of deionized water to the nested coffee cups and determine the total mass (cups + water); then assemble the calorimeter (add stir bar and lid).

34. Insert the thermometer in the calorimeter and clamp in place; be sure the thermometer bulb is immersed in the liquid and not touching the sides or bottom of the cup, and is safely above the stir bar. Record the temperature of the water (this will be the initial temperature of the water recorded in line #10, as well as for time = 0 temperature in line #12). Begin stirring.

35. Measure approximately 4 to 4.5 g of EITHER solid ammonium nitrate OR calcium chloride and record the mass precisely and the solid chosen. As both solids are hygroscopic, minimize the time the solid sample is exposed to the air.

36. Carefully add the entire solid sample to the calorimeter to avoid splashing and ensuring that the stirring is maintained. Replace the lid.

37. Record the time and temperature every 10 to 15 s for a total period of 8 to 10 min.

38. Once finished, dispose of the contents of the calorimeter as indicated by your instructor and rinse with water.

## I.  Thin Layer Chromatography

1.  TLC sketches (label each lane in each chromatogram with the substances spotted). Be sure to include a line for the solvent/eluent front and the spots observed under UV light.

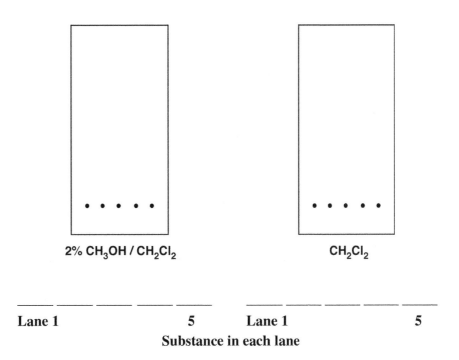

Substance in each lane

## II.  The Volatility of a Solution Containing a Nonvolatile Solute

2.  Time and mass measurements.

|  | Initial Time | Initial Mass (g) | Final Time | Final Mass (g) |
|---|---|---|---|---|
| methanol |  |  |  |  |
| 75% methanol |  |  |  |  |
| 50% methanol |  |  |  |  |
| 25% methanol |  |  |  |  |

## III. A.  Determination of the Heat Capacity (Calibration) of the Coffee-Cup Calorimeter

3.  Mass of two Styrofoam cups: _____

4.  Mass of cups + ~70 mL "cool water" (i.e., room temperature water): _____

5.  Mass of ~30 mL "hot water," $mass_{hot}$: _____

6.  Initial temperature of hot water (i.e., water in water bath), $T_{i, hot}$: _____

7.  Initial temperature of cool water, $T_{i, cool}$: _____

**8.** Temperature vs. time data:

| Time | Temperature (°C) | | Time | Temperature (°C) |
|---|---|---|---|---|
| 0 | | | | |
| | | | | |
| | | | | |
| | | | | |
| | | | | |
| | | | | |
| | | | | |
| | | | | |
| | | | | |
| | | | | |
| | | | | |
| | | | | |
| | | | | |
| | | | | |
| | | | | |
| | | | | |
| | | | | |
| | | | | |
| | | | | |

## III. B.  Determination of the Heat of Solution for _____ (list type of solid sample)

**9.** Mass of cups + ~100 mL water: _____

**10.** Initial temperature of water, $T_i$: _____

**11.** Mass of solid sample: _____

**12.** Temperature vs. time data:

| Time | Temperature (°C) | | Time | Temperature (°C) |
|---|---|---|---|---|
| 0 | | | | |
| | | | | |
| | | | | |
| | | | | |
| | | | | |
| | | | | |
| | | | | |
| | | | | |
| | | | | |
| | | | | |
| | | | | |
| | | | | |
| | | | | |
| | | | | |
| | | | | |
| | | | | |
| | | | | |
| | | | | |

| Time | Temperature (°C) | | Time | Temperature (°C) |
|---|---|---|---|---|
| | | | | |
| | | | | |
| | | | | |
| | | | | |
| | | | | |
| | | | | |
| | | | | |
| | | | | |
| | | | | |
| | | | | |
| | | | | |

## Analysis

### I.  Thin Layer Chromatography

**13.** Rank the five substances from most polar to least polar.

_____  _____  _____  _____  _____

    **MOST**                                         **LEAST**

### II.  The Volatility of a Solution Containing a Nonvolatile Solute

**14.** Summary data table

| | Evaporation Time (min) | Sample Mass Remaining (%) | Mole Fraction of Methanol |
|---|---|---|---|
| methanol | | | |
| 75% methanol | | | |
| 50% methanol | | | |
| 25% methanol | | | |
| ethylene glycol | _____ | 100 | 0 |

**15.** Sample calculation: Mole fraction of methanol in 75% methanol/25% ethylene glycol (v:v) (density of methanol = 0.7918 g mL$^{-1}$; density of ethylene glycol = 1.1132 g mL$^{-1}$)

**16.** Sample calculation: Sample mass remaining (%) for 75% methanol/25% ethylene glycol (v:v) solution
Sample mass remaining (%) = [final mass (g)/initial mass (g)] $\times$ 100

**17.** Plot of mole fraction of methanol vs. the percentage of the original sample mass remaining (%)

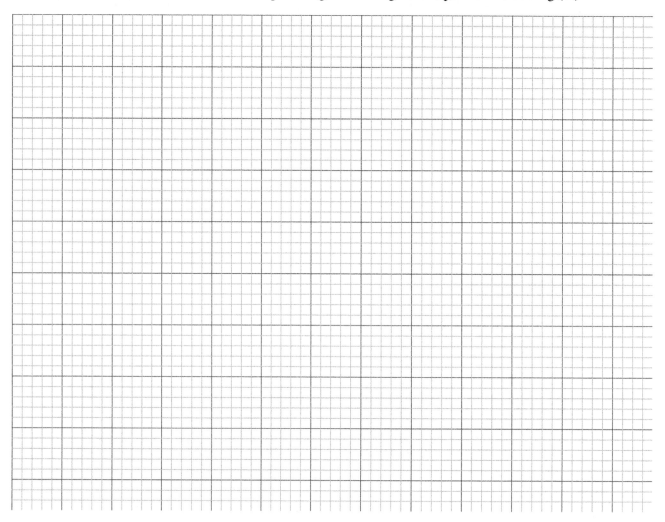

### III. A. <u>Determination of the Heat Capacity (Calibration) of the Coffee-Cup Calorimeter</u>

**18.** Mass of cool water: $mass_{cool} = mass_{cups + cool water} - mass_{cups}$

$$= (\#4) - (\#3) = \underline{\hspace{3cm}}$$

**19.** Maximum temperature: construct a plot of temperature vs. time using data from #8 (as shown in Figure 1 of Lab 12). From that, determine the maximum temperature (i.e., $T_f$)—this will be the $y$-intercept of the extrapolation line; the line can be obtained using Excel, a graphing calculator, or the least-squares method you used in Lab 4.

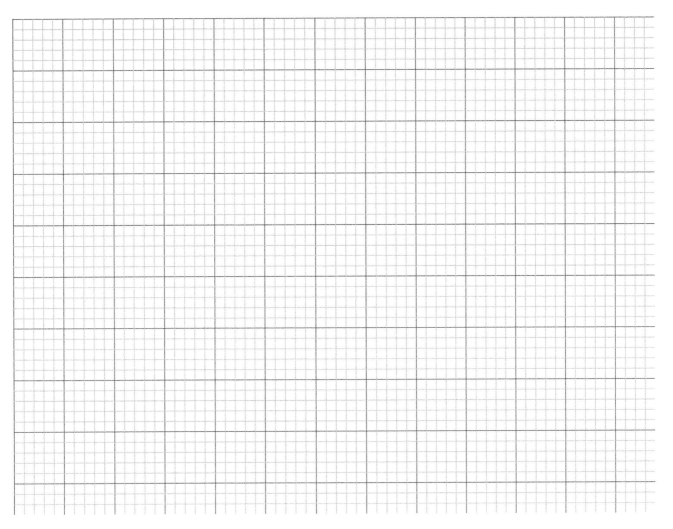

$$T_{\mathrm{f}} = \text{\underline{\hspace{3cm}}}$$

**20.** Heat exchanged between the hot and cool water:

$$-q_{\text{hot water}} = q_{\text{cool water}} + q_{\text{calorimeter}}$$

(a) $q_{hot\ water} = (s_{\text{water}}) \times (\text{mass}_{\text{hot}}) \times \Delta T$, where $\Delta T = T_{\mathrm{f}} - T_{\text{i, hot}}$

$= (4.184\ \mathrm{J\ g^{-1}\ {}^{\circ}C^{-1}}) \times (\#5) \times [(\#19) - (\#6)]$      $= \text{\underline{\hspace{2.5cm}}}$

(b) $q_{\text{cool water}} = (s_{\text{water}}) \times (\text{mass}_{\text{cool}}) \times \Delta T$, where $\Delta T = T_{\mathrm{f}} - T_{\text{i, cool}}$

$= (4.184\ \mathrm{J\ g^{-1}\ {}^{\circ}C^{-1}}) \times (\#18) \times [(\#19) - (\#7)]$      $= \text{\underline{\hspace{2.5cm}}}$

(c) $q_{\text{calorimeter}} = -(q_{\text{hot water}} + q_{\text{cool water}})$

$= -[(\#20a) + (\#20b)]$      $= \text{\underline{\hspace{2.5cm}}}$

**21.** Heat capacity of calorimeter:

$$q_{calorimeter} = (C_{calorimeter}) \times \Delta T$$

$$C_{calorimeter} = (\#20c)/[(\#19) - (\#7)]$$

= _____

## III. B. Determination of the Heat of Solution for _____ (list type of solid sample)

**22.** Mass of water:    $mass_{water} = mass_{cups + water} - mass_{cups}$

$$= (\#9) - (\#3)$$

= _____

**23.** Maximum or minimum temperature of solution: construct a plot of temperature vs. time using data from #12 (as shown in Figure 1 of Lab 12 for an exothermic process). From that, determine the maximum temperature if exothermic or minimum temperature if endothermic (i.e., $T_f$)—this will be the $y$-intercept of the extrapolation line.

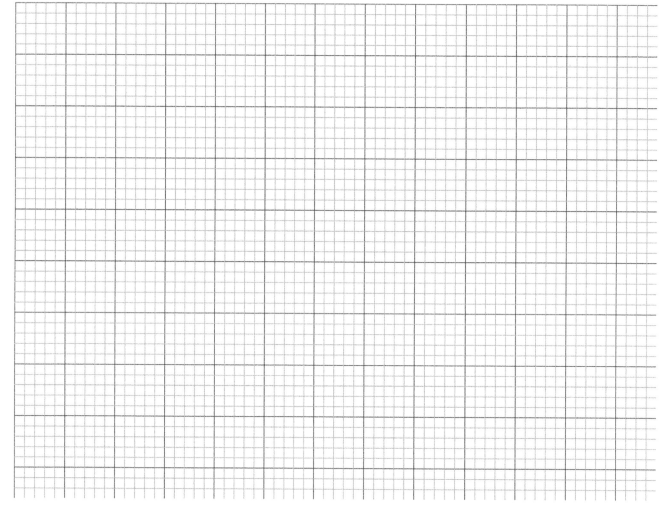

$T_f =$ _____

**24.** Heat absorbed or released during solution formation:

$$(a)\, q_{water} = (s_{water}) \times (mass_{water}) \times (T_f - T_{i\,water})$$

$$= (4.184\ \text{J g}^{-1}\,°\text{C}^{-1}) \times (\#22) \times [(\#23) - (\#10)]$$

= _____

(b) $q_{calorimeter} = (C_{calorimeter}) \times (T_f - T_{i\ water})$

$= (\#21) \times [(\#23) - (\#10)]$       = _____

(c) $q_1$, total heat absorbed or released by the surroundings (record the correct sign of this value)

$= (\#24a) + (\#24b)$       = _____

(d) $q_2$, heat released or absorbed during solution formation (record the correct sign of this value)

= _____

25. Moles of solid sample (from #11)       = _____

Show work:

26. Heat of solution, $\Delta H$:       $\Delta H = q_2/(\text{mol solid sample})$

$= (\#24d)/(\#25)$       = _____

## Reflection Questions

1. (a) Considering the five compounds that were studied by TLC (Figure 4), explain how their structures relate to their migration up the silica TLC plate.

(b) Which of the five compounds would you expect to have the lowest solubility in water?

(c) Choose one of the five compounds and provide a current use for it in society.

2.  (a) Draw Lewis structures for methanol, $CH_3OH$, and dichloromethane, $CH_2Cl_2$.

    (b) Methanol has a boiling point of 65°C and a molar mass of 32 g mol$^{-1}$. Dichloromethane has a boiling point of 40°C and a molar mass of 85 g mol$^{-1}$. Explain why the boiling point of methanol is significantly higher than that of dichloromethane despite dichloromethane being nearly three times as massive.

    (c) Based on your TLC results, what would you predict to happen if the same TLC experiment was performed with 5% methanol/dichloromethane as the eluent? Why?

3.  For the solution in Part II that was prepared from 50% methanol and 50% ethylene glycol *by volume*, calculate the following.

    (a) molality:

    (b) mass percent:

4. (a) In your plot of mole fraction of methanol vs. the percentage of the original sample mass remaining, a curved line was produced. If the solutions were ideal, Raoult's law would predict straight-line behavior. Provide an explanation for the observed deviation from Raoult's law. (Hint: Consider the structures of the two components that make up the solution and the type of deviation. Is the remaining sample mass for the three solutions greater or less than expected in comparison to those of the pure substances ethylene glycol and methanol?)

(b) Why was the percentage of the original sample mass remaining used for your plot and not simply the remaining mass in grams of the original sample?

5. Why was the *extrapolated* final temperature (and not the maximum or minimum *recorded* temperature) used to determine the heat of solution?

6. Look up the value for the heat of solution for $CaCl_2$ or $NH_4NO_3$ as appropriate and calculate your % error. Explain possible reasons for any discrepancies.

7. (a) Considering the energetics of solution formation and the solution you prepared in Part III of this lab, identify the attractive forces present in the pure substances and the solution.

Attractive forces in solute:

Attractive forces in solvent:

Attractive forces in solution:

(b) Comment on which attractive forces are greater, those in the pure substances or those in the solution? How do you know?

(c) What factor that does not relate to chemical structure promotes the formation of solutions?

## Connection

Based on your experience in this lab, draw a connection to something in your everyday life or the world around you (something not mentioned in the background section):

# 16

# Molar Mass Determination Through Freezing Point Depression

## Objectives

- Understand how colligative properties affect the freezing point of a liquid.
- Determine the molar mass of ethylene glycol using freezing point depression measurements.

## Materials Needed

- 600-mL beaker
- 400-mL beaker (2)
- 25 mm × 150 mm Pyrex culture tube
- Calibrated thermometer (or other device for monitoring temperature)
- Ring stand and clamp (appropriate size to secure culture tube)
- Thermometer clamp
- Stainless steel wire from which to fashion a stirring loop
- Stir rod
- Pasteur pipette
- Deionized water
- Ice
- Sodium chloride (NaCl)
- Ethylene glycol

## Safety Precautions

- Gloves and safety goggles/glasses should be worn during this lab.
- Ethylene glycol is moderately toxic if ingested. Avoid swallowing.

## Background

The addition of a nonvolatile solute to a solvent modifies many of the physical properties of the solvent. For some of these properties, such as freezing and boiling point changes, osmotic pressure, and vapor pressure depression, there is a general dependence on the *number* of solute particles added to the solvent (i.e., concentration), and *not* on the *identity* of the solute particles; these are known as **colligative properties.**

### FREEZING POINT DEPRESSION/BOILING POINT ELEVATION

The freezing point ($T_f$) of a liquid is the temperature at which the solid and liquid phases are in equilibrium. When a solution is freezing, the solute slows the formation of the solvent's solid lattice by physically impeding the condensation process, which then requires a lower temperature (i.e., more kinetic energy must be extracted) to get the solvent to begin freezing; this leads to a decrease in the freezing point. This effect can also be thought of in terms of a disruption of the solid/liquid equilibrium:

$$A(s) \xrightleftharpoons[\text{melting}]{\text{freezing}} A(l)$$

By decreasing the rate of freezing, but leaving the rate of melting unchanged, the solute causes a shift in the equilibrium to the right. Examples of this freezing point depression include the following:

- Antifreeze is used to lower the freezing point and raise the boiling point of your engine's coolant.
- Inorganic salts are commonly used to melt ice on roads ($CaCl_2$, $MgCl_2$, $KCl$, $NaCl$, $KC_2H_3O_2$).
- Interestingly, some animals that live in very cold locations have evolved to have high concentrations of inert substances (e.g., sorbitol or glycerol) dissolved in cellular fluids that depress the freezing point, protecting the organism from the low temperatures; these organisms include rainbow smelt and other artic fish, and the spring peeper frog (releases large amounts of glucose from its liver when subjected to cold temperatures).
- Once thought to be frozen solid, Lake Vida, an Antarctic lake, has recently been found to be in the liquid state beneath over 19 m of ice due to a concentration of salt that is seven times greater than typical seawater, despite a water temperature of $-10°C$![1]

We can see the decrease in freezing point experimentally by comparing the *cooling curves* (plot of temperature vs. time) for a pure substance and a solution (Figure 1). For a pure substance like water, the temperature decreases in a linear fashion as thermal energy is removed from the liquid (as it is placed into contact with something colder). In the absence of proper stirring, the liquid may eventually supercool—lowering of the temperature of a substance below its freezing point without solidification of the substance—because cooling is occurring faster than the molecules can lock into the solid lattice structure. Once a solid begins forming, the temperature returns to the freezing point and remains relatively constant until the whole sample is frozen. Then the temperature of the solid will begin decreasing.

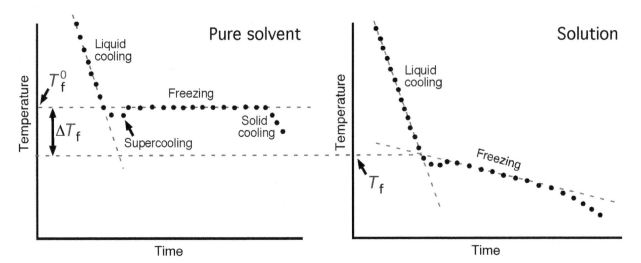

**Figure 1**  Cooling curves for a pure solvent (left) and a solution (right). On each graph: the data points represent the cooling curve; dotted lines highlight the liquid cooling region, the freezing region (i.e., liquid/solid phase change region), and extrapolations to obtain the freezing points ($T_f$ and $T_f^0$). $\Delta T_f^0$ represents the change in freezing point caused by addition of a solute to the pure solvent.

A similar curve can be generated for a solution (Figure 1). The major difference from the pure solvent cooling curve is that the "line" that represents the liquid/solid phase change is not a flat horizontal line; instead, the freezing region has a negative slope, indicating an ever-decreasing freezing point temperature. As the solution begins to freeze, the pure solvent forms crystals, while the solute stays dissolved in the remaining liquid (which then becomes more concentrated); the remaining solution is now more concentrated and therefore has a lower freezing point. This process continues, with the solvent molecules freezing out of solution, leaving behind a more concentrated solution with an even lower freezing point.

For each cooling curve, one can determine the freezing point by drawing two lines, one through the liquid cooling data points, and a second through the phase change region; the intersection of the two lines is the freezing point (extrapolate back to the $y$-axis, as shown in Figure 1). To accurately determine this point, first determine the equation for these two lines using (1) the method of least squares, (2) Excel, or (3) a graphing calculator. To then find the intersection of these two lines, solve the following two equations simultaneously for $y$ [both lines share one point $(x,y)$, which is the intersection point]:

$$y = m_1x + b_1 \qquad \text{Equation 1}$$

$$y = m_2x + b_2 \qquad \text{Equation 2}$$

Solve Equation 1 for $x$ and substitute that into Equation 2:

$$x = \frac{y - b_1}{m_1} \qquad \longrightarrow \qquad y = m_2\left(\frac{y - b_1}{m_1}\right) + b_2 \qquad \text{Equation 3}$$

Equation 3 can be rearranged, giving an equation which can be used to solve for $y$ (the freezing point, or $T_f$):

$$y = \frac{m_1b_2 - m_2b_1}{m_1 - m_2} \qquad \text{Equation 4}$$

The difference between the freezing points for the solution ($T_f$) and the pure solvent ($T_f^0$) is the freezing point depression ($\Delta T_f$):

$$\Delta T_f = T_f - T_f^0 \qquad \text{Equation 5}$$

The addition of a nonvolatile solute to a solvent will also have an effect on the boiling point. The boiling point is the temperature at which the liquid and gas phases of a substance are in equilibrium, and is also defined by the temperature at which the vapor pressure of the substance is equivalent to the atmospheric pressure. The vapor pressure of a volatile solvent is decreased when a nonvolatile solute is added due to the blockage of surface sites in the solution by the solute (Figure 2)—this lowers the number of surface sites from which the solvent molecules can escape into the vapor phase (see the Raoult's law discussion in your General Chemistry textbook). Since the vapor pressure is lower for the solution, more energy (i.e., a higher temperature) is required to reach the boiling point for the solution; in other words, the boiling point of the solution will be greater than the boiling point of the pure solvent.

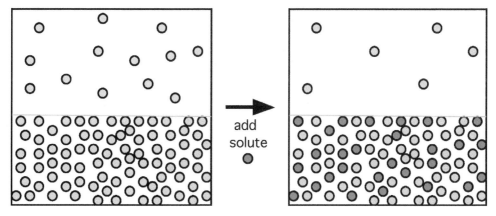

**Figure 2**  Nonvolatile solute particles occupy surface sites in the liquid phase of a solution, leading to fewer gas phase solvent particles and a lower vapor pressure for the solvent, the volatile component of the solution.

The changes in boiling point and freezing point for a solution are reflected in the phase diagram (Figure 3).

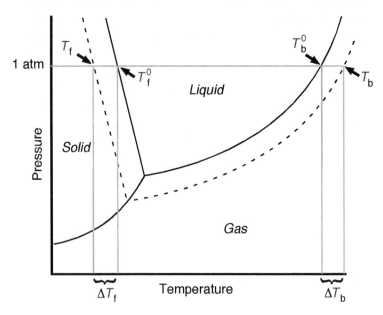

**Figure 3**   Phase diagram of water (solid lines/curves) and an aqueous solution (dotted line/curve). The normal freezing and boiling points ("normal" indicates at 1 atm pressure) are labeled.

The freezing and vaporization curves for the solution are shown as dotted lines, while those for the pure solvent are shown as solid lines. The freezing point of the solution is lowered (freezing curve shifts to the left) and the boiling point of the solution is elevated (vaporization curve shifts to the right). The area occupied by the liquid phase for the solution is therefore increased relative to the pure solvent.

The change in freezing point ($\Delta T_f$) and boiling point ($\Delta T_b$) have been shown experimentally to be directly proportional to the molal concentration of a dilute solution (Equation 6)

$$\Delta T_f = K_f \times m \qquad\qquad \Delta T_b = K_b \times m \qquad\qquad \textbf{Equation 6}$$

$$\Delta T_x = \left| T_x^0 - T_x \right| \qquad\qquad \textbf{Equation 7}$$

where $T_f^0$ and $T_b^0$ are the freezing point and boiling point of pure solvent, respectively, and $m$ is the molality of the solution. For freezing point depression, $\Delta T_f$ is subtracted from $T_f^0$; for boiling point elevation, $\Delta T_b$ is added to $T_b^0$. Molality is used since this concentration unit is insensitive to temperature fluctuations. $K_f$ and $K_b$ are the molal freezing-point depression constant and molal boiling-point elevation constant, respectively; these constants do not depend on the nature of the solute, but instead on the nature of the solvent. Table 1 contains a list of $K_f$ and $K_b$ values for different substances.

## ELECTROLYTE SOLUTIONS

As indicated before, the magnitude of change in a colligative property depends on the number of particles in solution (the molality, to be exact). For a nonelectrolyte, each mole of particles placed into solution yields 1 mole of dissolved particles. For a strong electrolyte (soluble salts, strong acids, and bases), however, each formula unit completely dissociates (or ionizes) to generate two or more dissolved particles.

$$NaCl(s) \xrightarrow{\sim 100\%} Na^+(aq) + Cl^-(aq)$$

$$Ca(NO_3)_2(s) \xrightarrow{\sim 100\%} Ca^{2+}(aq) + 2\,NO_3^-(aq)$$

$$HCl(aq) \xrightarrow{\sim 100\%} H^+(aq) + Cl^-(aq)$$

## Table 1 $K_f$ and $K_b$ Values for Different Substances.[2–5]

| Solvent | Normal Freezing Point (°C) | $K_f$ (°C $m^{-1}$) | Normal Boiling Point (°C) | $K_b$ (°C $m^{-1}$) |
|---|---|---|---|---|
| Water | 0.0 | 1.86 | 100.0 | 0.513 |
| Ethanol | −114.6 | 1.99 | 78.4 | 1.23 |
| t-Butanol | 25.5 | 9.10 | 82.0 | |
| Acetic acid | 16.6 | 3.90 | 118.1 | 3.22 |
| Benzene | 5.5 | 5.12 | 80.1 | 2.64 |
| Naphthalene | 80.2 | 6.8 | 217.9 | 5.8 |
| Camphor | 179 | 37.7 | 204.0 | 5.95 |
| Cyclohexane | 6.55 | 20.2 | 80.7 | 2.92 |

For a weak electrolyte (weak acids and bases), partial dissociation (or ionization) will generate more than the number of formula units dissolved, but less than would be expected if complete dissociation/ionization occurred.

$$HC_2H_3O_2(aq) \xrightarrow{<100\%} H^+(aq) + C_2H_3O_2^-(aq)$$

A few sample calculations shown in Table 2 illustrate the effect that dissociation/ionization has on the freezing point of water.

## Table 2 Theoretical and Actual Freezing Points for Electrolyte Solutions.

| Solution | Ions | Theoretical Freezing Point | Actual Freezing Point |
|---|---|---|---|
| 1.00 $m$ NaCl | $Na^+(aq)$, $Cl^-(aq)$ | $= 0°C − \Delta T_f$ <br> $= 0°C − (2.00\ m)(1.86)$ <br> $= −3.72°C$ | −3.37°C |
| 1.00 $m$ $(NH_4)_2SO_4$ | $2NH_4^+(aq)$, $SO_4^{2-}(aq)$ | $= 0°C − (3.00\ m)(1.86)$ <br> $= −5.58°C$ | −3.6°C |

The actual freezing point typically deviates from the theoretical freezing point due to partial reformation of cation/anion ion pairs, especially in concentrated solutions, which then gives an actual molality for the solution that is less than that predicted by assuming complete dissociation. Therefore, the total ion concentration in a 1.00 $m$ NaCl is greater than 1.00 $m$ but less than 2.00 $m$. The more dilute the solution, the closer the theoretical and actual freezing points will be. The **van't Hoff factor** ($i$) accounts for the effects of dissociation of a solute as it dissolves in a solvent, and can be calculated using Equation 8:

$$i = \frac{\text{actual number of particles in solution after dissociation}}{\text{number of formula units initially dissolved in solution}} \qquad \textbf{Equation 8}$$

We can include $i$ in our calculation of freezing point depression or boiling point elevation:

$$\Delta T_x = K_x \times (i \times m)$$

**Equation 9**

where "$i \times m$" is the actual molality of particles in solution. So, for the examples in Table 2, theoretical values for $i$ are calculated from the chemical formulas of the compounds [$i = 2$ for NaCl and 3 for $(NH_4)_2SO_4$]. However, due to ion pairing, the actual values for $i$ will be less than the theoretical values. The actual values of $i$ are calculated using Equation 9 and the experimental value of $\Delta T$ [$i = 1.91$ for NaCl and 1.94 for $(NH_4)_2SO_4$].

For a nonelectrolyte [e.g., sucrose and urea ($CH_4N_2O$)], $i = 1$.


## THIS LAB

The specific colligative property that will be studied in this lab is the freezing point depression of water upon addition of the solute ethylene glycol (a nonvolatile, nonelectrolyte). Ethylene glycol is a major component of the antifreeze used in automobiles to prevent overheating and freezing in the engine. It had no commercial use before World War I, and was then primarily used in dynamite manufacture for many years. Besides antifreeze, current uses include hydraulic brake fluid; a deicing agent for airport runways, aircraft, and boats; a minor component in shoe polish; and a preservative of biological samples (a safer alternative to formaldehyde). Interestingly enough, it has also been observed in outer space.[6]

You will use $\Delta T_f$ to determine the molecular mass of ethylene glycol. From the measurement of $\Delta T_f$, the molality of the solution can be determined from:

$$m = \frac{\Delta T_f}{K_f}$$

**Equation 10**

Using the definition of molality and a known mass of solvent, the moles of ethylene glycol used can then be determined:

$$moles_{ethylene\ glycol} = m \times (kg\ H_2O)$$

**Equation 11**

Finally, the molar mass of ethylene glycol can be calculated as shown:

$$molar\ mass_{ethylene\ glycol} = \frac{grams_{ethylene\ glycol}}{moles_{ethylene\ glycol}}$$

**Equation 12**

The temperature at which a solution freezes is difficult to determine by simply observing the change in the liquid state for two reasons: (1) supercooling and (2) freezing of a solution occurs over a broad temperature range. $\Delta T_f$ will therefore be determined by measuring the cooling curves for pure water and an ethylene glycol/water solution, as shown in Figure 1. Finally, we will use an ice/salt water bath to generate temperatures at or below $-10°C$ to create the cooling curves (interestingly enough, the freezing point depression of the ice water caused by the added salt is needed to get these temperatures!).


## *References*

1. Doran PT, Fritsen CH, McKay CP, Priscu JC, Adams EE. "Formation and Character of an Ancient 19-M Ice Cover and Underlying Trapped Brine in an 'Ice-Sealed' East Antarctic Lake." *Proc. Natl. Acad. Sci. U.S.A.* 2003; 100: 26–31.
2. Eastman ED, Rollefson GK. (1947). *Physical Chemistry.* McGraw-Hill, New York.
3. Pauling L. (1970). *General Chemistry.* Dover Publications, New York.
4. Moore WJ. (1962). *Physical Chemistry.* Prentice Hall, Englewood Cliffs, NJ.
5. Lide DR, Ed. (2003). *CRC Handbook of Chemistry and Physics.* 84 ed. CRC Press, Boca Ratan, FL.
6. Hollis JM, Lovas FJ, Jewell PR, Coudert LH. "Interstellar Antifreeze: Ethylene Glycol." *The AstroPhysical Journal.* 2002; 571: L59–L62.

## Procedure

### I. Prepare ice/saltwater bath

1. Determine the mass of NaCl necessary to lower freezing point of 300 mL of distilled water to $-10°C$ (perform calculations in the "Analysis" section).
2. Measure out the mass of NaCl that you determined and record the actual mass.
3. Carefully place a 400-mL beaker inside a 600-mL beaker (outer beaker will serve as an added layer of insulation).
4. To create the ice/saltwater bath, fill the 400-mL beaker to the 250 to 270 mL mark with ice, pour the NaCl over the ice, and then pour approximately 30 to 50 mL of deionized water over the ice/salt. Stir the mixture thoroughly with a glass rod to form a slush.

    **Caution:** Do not stir the ice/saltwater bath with a thermometer. Use a stir rod.

5. Measure the temperature of the ice/saltwater mixture; once the temperature reaches a constant value, record the temperature. Then dry the thermometer; use a thermometer clamp to anchor the thermometer above the ice/saltwater bath in preparation to measure cooling curves. Allow the temperature of the thermometer to reach room temperature before continuing.

*Note:* This ice/saltwater bath will be used through the rest of the lab to generate cooling curves of both water and aqueous ethylene glycol solutions. Before each experiment, be sure to check the temperature of the bath. If the temperature rises above $-5°C$, then pour off the excess water and add more ice and salt until the temperature again drops to approximately $-10°C$.

### II. Molar mass determination of ethylene glycol: cooling curve for pure water

6. Take the stainless steel wire and fashion it into a wire stirrer, as shown in Figure 4. The loop on one end should be small enough to fit into the culture tube and large enough to fit around the thermometer; this should allow stirring of a liquid in the tube through an up-and-down motion of the wire stirrer.

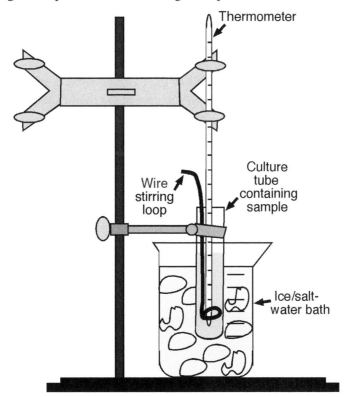

**Figure 4 Experimental setup (see text for details).**

7. Place a clean and dry culture tube in a 250-mL beaker and measure and record the mass. Then add approximately 20 mL of deionized water (no need to record the exact mass of the water at this time).

8. Measure and record the original temperature of the deionized water. This temperature should be recorded in the time = 0:00 entry in the "Water Cooling Curves" table (#4 in Data Collection section).

9. *This next step should be accomplished as quickly as possible so that temperature vs. time data can be collected.* Place the culture tube into the ice/saltwater bath (the level of the liquid in the tube should be below the ice level) and secure in place with the ring stand clamp. Insert the wire stirrer and then lower the thermometer into the culture tube through the loop of the stirrer; be sure the bulb of the thermometer is placed in the middle of the liquid. The completed assembly should look like Figure 4.

10. While constantly stirring, record the temperature of the liquid every 5 to 30 s for approximately 5 min until the temperature becomes stable and you collect 5 to 10 measurements with approximately the same temperature (the time interval used will be based on the student's ability to accurately record time/temperature data—if working alone, then up to 30 s might be necessary, while student pairs can record data faster; if using an automated temperature recorder, then 5 s intervals will produce a very nice plot).

11. Bring the temperature of the liquid in the tube up to room temperature (this may take several minutes—it can be expedited by warming the tube in your hands).

12. Repeat steps 7 to 10 once more to have two complete cooling curves for water. DO NOT DISPOSE OF THE WATER.

### III. Molar mass determination of ethylene glycol: cooling curve for aqueous ethylene glycol solution

13. Clean and dry the outside of the culture tube thoroughly, then place it into the same 250-mL beaker and determine the mass again using the same balance as before—this will be the mass of the beaker + culture tube + water.

14. Add ~2.0 g of ethylene glycol (density = 1.113 g mL$^{-1}$) to the tube—this can be accomplished using a Pasteur pipette; record the mass.

15. Mix the two substances thoroughly by tapping the outside of the tube (the mixing loop can also be used to aid in mixing).

16. Measure and record the original temperature of the mixture. This temperature should be recorded in the time = 0:00 entry in the "Ethylene Glycol/Water Cooling Curves" table (#7 in Data Collection section).

17. Place the tube into the ice/saltwater bath and assemble the apparatus as you did in step 9.

18. While constantly stirring, record the temperature of the solution every 5 to 30 s for approximately 8 min until you collect *at least* 10 measurements below the freezing point of pure water.

19. Remove the culture tube from the bath and place the contents in the marked waste container provided by your instructor. Rinse and dry the inside and outside of the tube.

20. Place the clean and dry tube into the 250-mL beaker and again determine their combined mass (record). Add approximately 20.0 mL of distilled water to the tube.

21. Repeat steps 13 to 18 once more to have two complete cooling curves for the solution.

## Data Collection

### I. Prepare ice/saltwater bath

1. Mass NaCl: _____
2. Temperature of ice/saltwater bath: _____

### II. Molar mass determination of ethylene glycol: cooling curve for pure water

3. Mass of beaker + culture tube: _____

**4.** Cooling curve temperature data:

| | Water Cooling Curves | |
| --- | --- | --- |
| | Trial 1 | Trial 2 |
| Time | Temperature (°C) | Temperature (°C) |
| | | |
| | | |
| | | |
| | | |
| | | |
| | | |
| | | |
| | | |
| | | |
| | | |
| | | |
| | | |
| | | |
| | | |
| | | |
| | | |
| | | |
| | | |
| | | |
| | | |
| | | |
| | | |
| | | |
| | | |
| | | |
| | | |
| | | |
| | | |
| | | |
| | | |
| | | |
| | | |

### III. Molar mass determination of ethylene glycol: cooling curve for aqueous ethylene glycol solution

5. Mass of beaker + culture tube + water:

Trial 1: _____

Trial 2: _____

6. Mass of #5 + ethylene glycol:

Trial 1: _____

Trial 2: _____

7. Cooling curve temperature data:

| Ethylene Glycol/Water Cooling Curves | | |
|---|---|---|
| | Trial 1 | Trial 2 |
| Time | Temperature (°C) | Temperature (°C) |
| | | |
| | | |
| | | |
| | | |
| | | |
| | | |
| | | |
| | | |
| | | |
| | | |
| | | |
| | | |
| | | |
| | | |
| | | |
| | | |
| | | |
| | | |
| | | |
| | | |
| | | |
| | | |
| | | |
| | | |
| | | |
| | | |
| | | |
| | | |
| | | |
| | | |
| | | |

### I. Prepare ice/saltwater bath

**8.** Determine the mass of NaCl necessary to lower freezing point of 300 mL of distilled water to $-10°C$:

Mass NaCl needed: _____

**9.** Based on the actual mass of NaCl you used (#1), determine the theoretical freezing point of the ice/saltwater bath:

Theoretical freezing point: _____

### II. Molar mass determination of ethylene glycol: cooling curve for pure water

**10.** Plot cooling curve (temperature vs. time) for water (data in #4):

**Trial 1:**

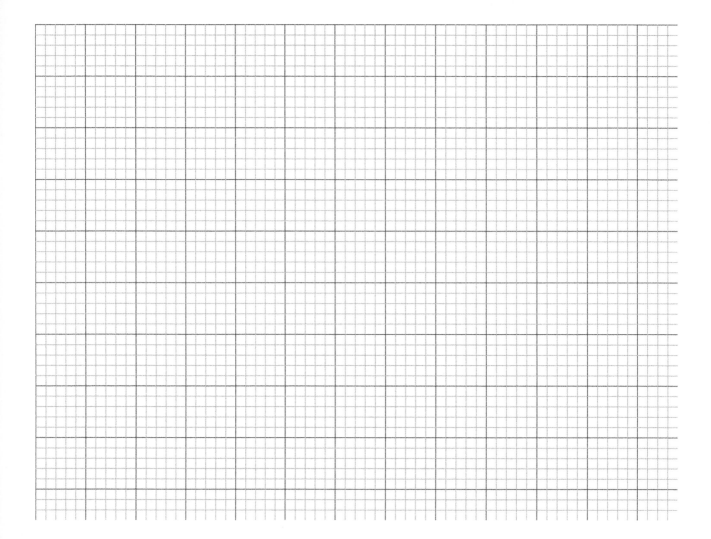

**Trial 2:**

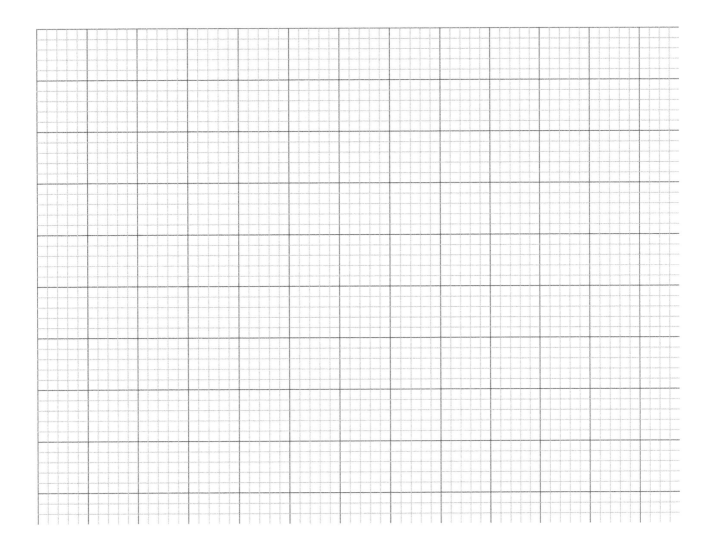

Laboratory 16   Molar Mass Determination Through Freezing Point Depression

11. Use a straightedge (e.g., a ruler) to draw two straight lines on each cooling curve—one drawn through the data points before the water freezes, and one drawn through the data points that make up the region after the water begins freezing (e.g., the region where the curve plateaus). The intersection of these two lines is the freezing point (determine by extrapolating this point to the *y*-axis).

Freezing point of water:     Trial 1: _____

Trial 2: _____

Average: _____

12. Determine the % error for the freezing point of water (comparing the average temperature you determined in #11 and the actual freezing point; *Note:* use Kelvin):

% Error: _____

### III. Molar mass determination of ethylene glycol: cooling curve for aqueous ethylene glycol solution

13. Mass of water:     Trial 1: _____

Trial 2: _____

14. Molality of aqueous ethylene glycol solution:     Trial 1: _____

Trial 2: _____

15. Plot cooling curve (temperature vs. time) for water (data in #7):

**Trial 1:**

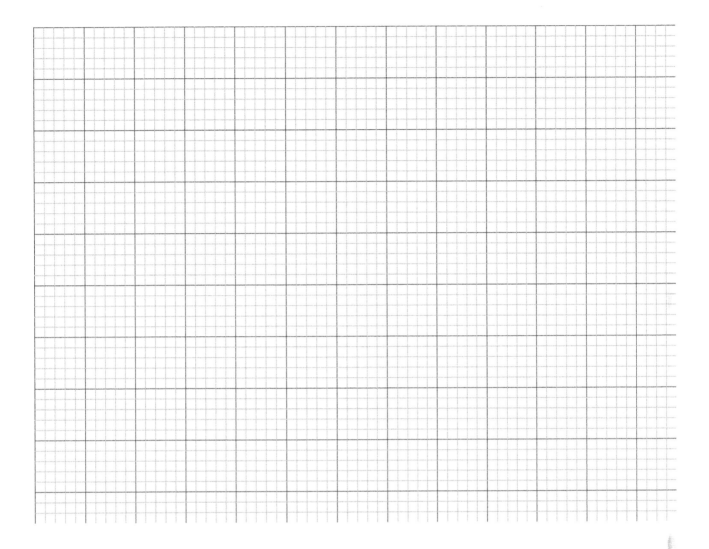

**Trial 2:**

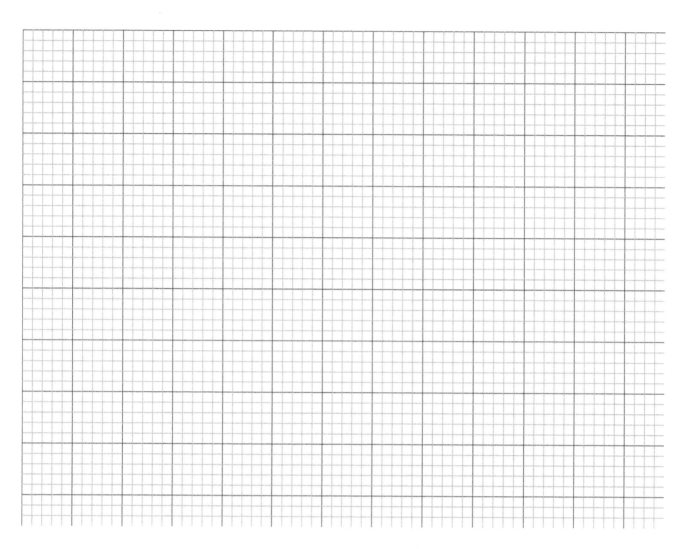

16. Use a straightedge (e.g., a ruler) to draw two straight lines on each cooling curve—one drawn through the data points before the solution begins to freeze, and one drawn through the data points that make up the region after the solution begins freezing. The intersection of these two lines is the freezing point (determine by extrapolating this point to the y-axis).

    Freezing point of solution:

    Trial 1: _____

    Trial 2: _____

    Average: _____

17. Determination of molar mass of ethylene glycol calculations:

    $$\Delta T_f = \left| T_{f,\,\text{solution}} - T_{f,\,\text{water}} \right| = \left| (\#16) - (\#11)_{\text{average}} \right| =$$

    $\Delta T_f$ (Trial 1): _____

    $\Delta T_f$ (Trial 2): _____

$$\Delta T_f = (\text{molality}_{\text{solution}}) \times (1.86°C \ m^{-1})$$

molality solution (Trial 1) = _____

molality solution (Trial 2) = _____

$$\text{molality}_{\text{solution}} = \frac{\text{mol}_{\text{ethylene glycol}}}{\text{kg water}}$$

$\text{mol}_{\text{ethylene glycol}}$ (Trial 1) = _____

$\text{mol}_{\text{ethylene glycol}}$ (Trial 2) = _____

$$\text{mol}_{\text{ethylene glycol}} = \frac{\text{mass}_{\text{ethylene glycol}}}{\text{molar mass}_{\text{ethylene glycol}}}$$

molar mass$_{\text{ethylene glycol}}$ (Trial 1) = _____

molar mass$_{\text{ethylene glycol}}$ (Trial 2) = _____

Average molar mass$_{\text{ethylene glycol}}$ = _____

18. The actual molar mass of ethylene glycol is 62.07 g mol$^{-1}$. Calculate the % error for your average molar mass:

% Error: _____

## Reflection Questions

1. If you assume that *no* experimental errors were made when you made your ice/saltwater bath, determine the actual van't Hoff factor for your salt water sample:

2. Provide possible experimental errors to explain the % error in your $T_f$ of water (#12):

3. The structural formula of ethylene glycol is $CH_2OHCH_2OH$.
   (a) Draw the Lewis structure of ethylene glycol:

   (b) Explain why ethylene glycol is completely miscible with water (i.e., can be mixed in any proportions):

4. Provide possible experimental errors to explain the % error in your calculated molar mass of ethylene glycol (#18):

5. A thermometer is calibrated incorrectly to read 0.3°C higher than the actual temperature over its entire temperature range. If you had used this thermometer in this lab, how would it affect your calculated molar mass of ethylene glycol?

6. If the freezing point of your solution had been incorrectly read 0.4°C lower than the true freezing point (the freezing point of pure water was read correctly, however), would the calculated molar mass of the solute be too high or too low? Explain your answer.

7. In very cold areas, sand is often used instead of inorganic salts to battle icy roads.
   (a) Why might use of inorganic salts *not* be totally effective in very cold locations?

   (b) How does sand function to battle ice on roads?

## Connection

Based on your experience in this lab, draw a connection to something in your everyday life or the world around you (something not mentioned in the background section):

# 17

# Kinetics—The Hydrolysis of *p*-Nitrophenyl Acetate

## Objectives

- Determine the rate law for the hydrolysis of *p*-nitrophenyl acetate.
- Use integrated rate laws to determine the order of the reaction.
- Determine the rate constant for the reaction at room temperature.

## Materials Needed

- *p*-Nitrophenyl acetate (or 4-nitrophenyl acetate)
- *p*-Nitrophenol (or 4-nitrophenol)
- Methanol
- Deionized water*
- Standardized 0.0010 *M* sodium hydroxide (NaOH) solution (pH 11)
- Erlenmeyer flasks of various sizes for preparation of solutions
- Volumetric flasks
- 50- and 10-mL graduated cylinders
- Small test tubes
- Pipetter or Pasteur Pipette
- Method for accurately measuring the absorbance of samples at a wavelength of 405 nm. Possible options include:
  - Vernier LabPro handheld device with a SpectroVis Plus spectrophotometer attachment. The associated software LoggerPro must be installed on a computer to allow downloading and manipulation of data. Information about the Vernier device and associated peripherals and software can be obtained from the website http://www.vernier.com (last accessed May 2011).
  - UV/Vis spectrophotometer
  - Spectronic 20 spectrophotometer
- Cuvettes compatible with the device used to measure absorbance

## Safety Precautions

- Gloves and safety goggles/glasses should be worn during this lab.
- NaOH is a strong base that can cause burns. Avoid contact with body tissues.

---

*For best results, the deionized water should be boiled for 5 min to minimize the concentration of dissolved carbon dioxide. Basic solutions absorb and react with carbon dioxide from the air, which reduces the concentration of hydroxide in solution over time. Upon cooling, the boiled, deionized water should be kept in a sealed container and used throughout this lab whenever deionized water is needed.*

- Methanol is toxic if ingested or inhaled. It is also very flammable. Do not use methanol near open flames.
- Dispose of waste solutions in designated waste containers as indicated by your instructor.

## *Background*

Chemical reactions occur at different speeds. Some are fast, such as the ignition of gasoline fumes in an engine, while others are much slower, such as the hardening of concrete (regular concrete does not reach complete hardness for several years) and the conversion of diamond to graphite (millions of years). The speed of a reaction is known as the **reaction rate,** and the study of reaction rates is known as **kinetics.** Understanding reaction rates is important in a number of ways, from identifying methods to speed up desired reactions to minimizing damage caused by undesirable reactions to understanding the individual steps that convert reactants to products.

The rate of a reaction is commonly expressed as a change in the concentration of a reactant or product over a specific time interval. For example, given the reaction

$$A + B \rightarrow C$$

the reaction rate is expressed as

$$\text{rate} = -\frac{\Delta[A]}{\Delta t} = -\frac{\Delta[B]}{\Delta t} = \frac{\Delta[C]}{\Delta t} \qquad \textbf{Equation 1}$$

Because reaction rates (with units of $M\ s^{-1}$) are positive values and reactant concentrations decrease as the reaction proceeds, $\Delta[A]$ and $\Delta[B]$ in Equation 1 are multiplied by $-1$. One can report on the reaction rate by monitoring the concentration of any reactant or product through observation of a characteristic color change, conductance change, pressure change, and the like.

At the molecular level, most reactions occur through collisions between reactants, with the vast majority of collisions involving only two particles. The rate of a reaction is proportional to the number of effective collisions per second:

$$\text{rate } \alpha \ \frac{\#\text{ effective collisions}}{\text{second}} \qquad \textbf{Equation 2}$$

An effective collision is defined as one that yields product; only a small fraction of all collisions are effective. As moving particles collide, kinetic energy is transferred into vibrational energy. If the vibrational energy is large enough, bonds within the colliding molecules can break (the first step to forming products). There is a minimum energy required to initiate a reaction—this is known as the **activation energy** ($E_a$). If the collision energy is less than $E_a$, then no reaction occurs. An effective collision also requires an appropriate orientation between the colliding molecules; an incorrect orientation leads to an ineffective collision. With these two requirements, only a very small number of all collisions are effective. The number of effective collisions per unit time (and therefore the rate) can be increased by increasing the concentrations of the colliding species, and by increasing the average kinetic energy (i.e., the temperature) of the reaction.

## RELATIONSHIP BETWEEN REACTION RATE AND CONCENTRATION

In general, the dependence of the reaction rate on the concentration of reacting species can be expressed with an empirical equation known as the **rate law equation.** For the reaction

$$a\,A + b\,B \rightarrow c\,C \qquad \textbf{Reaction 1}$$

the rate law equation is written as in Equation 3:

$$\text{rate} = k[A]^x[B]^y \qquad \textbf{Equation 3}$$

[A] and [B] represent the molar concentrations of the reactants in the reaction (reaction rates are predominately dependent on the reactant concentrations at the beginning of the reaction where very little to no product concentration is present, so products are not included in the rate law); $k$ is the **rate constant** for the reaction, a proportionality constant whose value depends on temperature; and $x$ and $y$ are the **orders** of the reaction with respect to A and B, respectively. Reaction orders are commonly positive whole numbers. When the exponent $= 1$, it is **first order,** meaning the rate is directly proportional to the concentration of that species (i.e., if $x = 1$ in Equation 3, doubling [A] leads to a doubled rate); if the exponent $= 2$, it is **second order,** meaning the rate is proportional to the square of the concentration (i.e., if $x = 2$, doubling [A] leads to an increase in rate by $2^2$, or a factor of 4); if the exponent $= 0$, it is **zeroth order,** meaning that the rate does not depend on that concentration (i.e., if $x = 0$, doubling [A] leads to no change in the rate).

If Reaction 1 describes a single collision between two atomic/molecular species, then $x$ and $y$ are equal to the stoichiometric coefficients ($a$ and $b$) from the balanced reaction. *However, if the reaction instead describes a macroscopic chemical reaction as described by a standard overall balanced chemical equation, then x and y are NOT related to the coefficients and must be determined through experimentation;* this is because many macroscopic reactions actually occur through multiple steps (the series of steps is called the **reaction mechanism**). The balanced reaction then does not describe the details of these steps. In this case, to determine the dependence of the rate on each reactant in the chemical equation, one needs to perform a series of experiments.

A straightforward method for determining the values for the reaction orders and the rate constant in the rate law equation is the **method of initial rates.** In this method, a series of experiments are performed. First, the **initial rate** of the reaction (i.e., the instantaneous rate close to the beginning of the reaction) is measured for a set of initial reactant concentrations. Then the initial concentration of *one* of the species is changed (e.g., doubled), and the initial rate is again measured. A comparison of how the initial rate changes as one reactant concentration is changed allows the order of the reaction with respect to that reactant to be determined. Additional experiments are then conducted (each changing only one reactant concentration) until all exponents in the rate law equation are determined. Finally, to determine the value for the rate constant, the values for the initial rate and initial reactant concentrations from one of the experiments are substituted into the completed rate law equation, and the equation is solved for $k$ (the units on $k$ are important; they will vary depending on the values of the exponents). As a simple example, for Reaction 2, two experiments are run (Table 1):

$$2A \rightarrow C \qquad \textbf{Reaction 2}$$
$$\text{rate} = k[A]^x \qquad \textbf{Equation 4}$$

| Table 1   Kinetics data collected for Reaction 2. | | |
|---|---|---|
| Experiment | $[A]_0$ (*M*) | Initial rate (*M* s$^{-1}$) |
| 1 | 0.10 | $8.0 \times 10^{-4}$ |
| 2 | 0.20 | $1.6 \times 10^{-3}$ |

By inspection of the data, as $[A]_0$ is doubled, one sees that the initial rate of the reaction also doubles—this indicates that the exponent $x = 1$ (rate proportional to [A]) and the reaction is first order with respect to [A]. One can also get the value of $x$ through a simple algebraic relationship (which will be useful in situations where the numbers aren't quite so obvious): write a fraction where the numerator is the rate law equation with the values from one experiment filled in, and the denominator is the rate law equation with the values from the other experiment; then solve for $x$:

$$\frac{\text{rate}_{\text{exp 2}} = 1.6 \times 10^{-3} \, M \, \text{s}^{-1} = k[0.20 \, M]^x}{\text{rate}_{\text{exp 1}} = 8.0 \times 10^{-4} \, M \, \text{s}^{-1} = k[0.10 \, M]^x} \qquad \textbf{Equation 5}$$

$$\left(\frac{1.6 \times 10^{-3} \, M \, \text{s}^{-1}}{8.0 \times 10^{-4} \, M \, \text{s}^{-1}}\right) = \frac{k}{k}\left(\frac{[0.20 \, M]}{[0.10 \, M]}\right)^x \qquad \textbf{Equation 6}$$

$$(2) = \frac{k}{k}(2)^x \quad \rightarrow \quad x = 1$$

We could write Equations 5 and 6 with experiment 2 data in the denominator and experiment 1 data in the numerator, and this would yield the same answer. Finally, to solve for $k$, we substitute experiment 1 values into the rate law:

$$\text{rate} = 8.0 \times 10^{-4} \, M \, \text{s}^{-1} = k(0.10 \, M)$$

$$k = 8.0 \times 10^{-3} \, \text{s}^{-1} \qquad \text{(note the units)}$$

## RELATIONSHIP BETWEEN CONCENTRATION AND TIME

It is particularly useful to relate **concentration** to **time** (i.e., starting with $[X] = 0.100 \, M$, how long will it take until $[X]$ drops to $0.001 \, M$?). We are often more interested in the answer to the question "how long will it take for a reaction/process to be finished?" or "how much reactant is left after a certain period of time" vs. "how fast is the process proceeding right now?" To answer concentration vs. time questions, it is necessary to use an integrated form of the rate law equation. Fortunately, the same time dependence is observed for all reactions with the same reaction order.

### First-Order Reactions

By definition, the rate of first-order reactions depends on the concentration of a reactant raised to the first power. For the general Reaction 3, the rate law is then written as shown with rate defined as the change in the concentration of reactant over time (Equation 7).

$$a \, \text{A} \rightarrow \text{products} \qquad \textbf{Reaction 3}$$

$$\text{rate} = k[\text{A}] \quad \text{with} \quad \text{rate} = -\frac{\Delta[\text{A}]}{\Delta t} \qquad \textbf{Equation 7}$$

Calculus is then used to produce the integrated form of the rate law (Equation 12) as an equation that relates concentration of reactant to time:

$$\text{rate} = -\frac{d[\text{A}]}{dt} = k[\text{A}] \qquad \textbf{Equation 8}$$

$$\frac{1}{[\text{A}]}d[\text{A}] = -k \, dt \qquad \textbf{Equation 9}$$

$$\int \frac{1}{[\text{A}]}d[\text{A}] = -\int k \, dt \qquad \textbf{Equation 10}$$

$$\ln[\text{A}]_t = -kt + \text{C} \qquad \textbf{Equation 11}$$

$$\ln[A]_t = -kt + \ln[A]_0 \quad \text{or} \quad \ln\left(\frac{[A]_0}{[A]_t}\right) = kt \qquad \textbf{Equation 12}$$

Thus, Equation 12 is the **integrated rate law** for a first-order reaction—it relates $[A]$ at time $= 0$ and time $= t$. The equation is in the form of a straight line ($y = mx + b$); a plot of $\ln[A]$ vs. time generates a straight line with a slope $= -k$ and $y$-intercept of $\ln[A]_0$ (Figure 1).

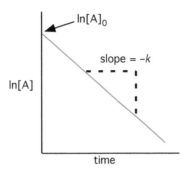

**Figure 1**   Plot of ln[A] vs. time for a first-order reaction.

## Second-Order Reactions

For second-order reactions, the rate depends on the concentration of a reactant raised to the second power. Using the same method as shown above for first-order reactions, an integrated rate law is derived (Equation 14):

$$a\,A \rightarrow \text{product}$$

$$\text{rate} = k[A]^2 \quad \text{with} \quad \text{rate} = -\frac{\Delta[A]}{\Delta t} \qquad \textbf{Equation 13}$$

$$\frac{1}{[A]_t} = kt + \frac{1}{[A]_0} \qquad \textbf{Equation 14}$$

Equation 14 is the integrated rate law for a second-order reaction. A plot of $1/[A]$ vs. time yields a straight line with a slope $= k$ and $y$-intercept of $1/[A]_0$ (Figure 2).

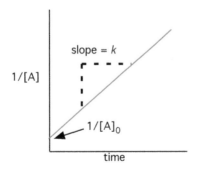

**Figure 2**   Plot of 1/[A] vs. time for a second-order reaction.

## Zeroth-Order Reactions

For zeroth-order reactions, the rate is independent of reactant concentration.

$$a\,A \rightarrow \text{product}$$

$$\text{rate} = k[A]^0 = k \quad \text{with} \quad \text{rate} = -\frac{\Delta[A]}{\Delta t}$$

$$[A]_t = -kt + [A]_0 \qquad\qquad \textbf{Equation 15}$$

Equation 15 is the integrated rate law for a zeroth-order reaction. A plot of [A] vs. time yields a straight line with a slope $= -k$ and $y$-intercept of $[A]_0$ (Figure 3).

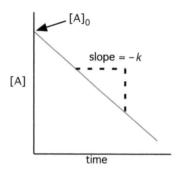

**Figure 3**  Plot of [A] vs. time for a zeroth-order reaction.

The order of a reaction can be graphically determined by measuring [A] vs. time. Then, plots of [A] vs. time, ln[A] vs. time, and 1/[A] vs. time are compared with the straight-line plot revealing the order of the reaction (0, 1, or 2). In addition, the slope of the straight-line plot gives the rate constant (Figure 4).

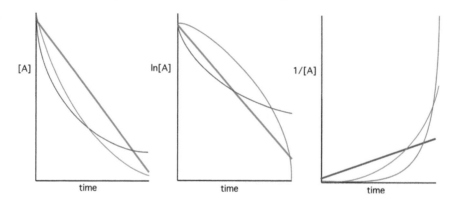

**Figure 4**  The order of a reaction with respect to [A] can be determined graphically. In each graph, the red plot is for a zeroth-order reaction, the green plot is for a first-order reaction, and the blue plot is for a second-order reaction. (Left) [A] vs. time plots—the straight line is for a zeroth-order reaction. (Middle) ln[A] vs. time plots—the straight line is for a first-order reaction. (Right) 1/[A] vs. time plots—the straight line is for a second-order reaction.

## PSEUDO-ORDER CONDITIONS

Multiple reactants are often present in chemical reactions with consequent multiple reactant concentrations necessarily included in the rate law (Equation 16):

$$a\,A + b\,B + c\,C \rightarrow \text{Products} \qquad\qquad \textbf{Reaction 4}$$

$$\text{rate} = k[A]^x[B]^y[C]^z \qquad\qquad \textbf{Equation 16}$$

To determine the value of the exponents $x$, $y$, and $z$, it is useful to conduct a series of experiments in which only one reactant concentration is changing with time; this allows one to determine how the rate is affected by a single species. One way to establish such conditions is if the initial concentrations of all but one species are significantly greater than the reactant of interest. So in the case of Reaction 4:

$$[B]_0 \text{ and } [C]_0 \gg [A]_0 \qquad \text{(therefore, [B] and [C] approximately constant)}$$

$$\text{rate} = k[A]^x[B]^y[C]^z = (k[B]^y[C]^z)[A]^x = k'[A]^x \qquad \textbf{Equation 17}$$

The constant $k'$ is known as a **pseudo-rate constant.** One can now collect [A] vs. time data and determine the order of the reaction with respect to [A] (i.e., the value of $x$ in Equation 16). The values of $y$ and $z$ in Equation 16 can then be determined in a similar fashion (e.g., set up conditions where $[A]_0$ and $[B]_0 \gg [C]_0$ so that the value of $z$ can be determined).

## THIS LAB

In this lab, the kinetics for the hydrolysis reaction of $p$-nitrophenyl acetate (or 4-nitrophenyl acetate) under basic conditions (i.e., in the presence of by $OH^-$) will be studied (Reaction 5):

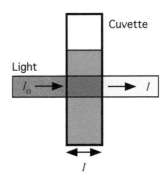

$$+ \; OH^- \longrightarrow \qquad + \; CH_3CO_2H \qquad \textbf{Reaction 5}$$

$p$-Nitrophenyl acetate (molecular formula = $C_8H_7NO_4$; molar mass = 181.1 g mol$^{-1}$; abbreviated NPA in this lab) is an ester—an organic molecule that contains the structure $R-(C=O)-O-R'$, where R and R' represent structural variations that lead to a wide range of known esters. NPA is a common substrate used to study esterases, which are enzymes that catalyze the breakdown of esters.[1–4] A hydrolysis reaction involves the reaction of water with a substrate, a primary way that organic esters are broken down in aqueous environments. The hydrolysis rate of NPA is increased in the presence of $OH^-$.

The rate of the reaction between NPA and $OH^-$ is given by the generalized rate law Equation 18:

$$\text{rate} = k[NPA]^x[OH^-]^y \qquad \textbf{Equation 18}$$

where $k$ is the rate constant for the reaction, $x$ is the reaction order with respect to NPA, and $y$ is the reaction order with respect to $OH^-$. The values of $x$ and $y$ will be determined experimentally. The $[OH^-]_0$ will be made significantly larger than $[NPA]_0$; therefore, $[OH^-]^y$ will not change appreciably during the course of the reaction and can be considered constant. The rate law therefore simplifies to

$$\text{rate} = k'[NPA]^x \qquad \textbf{Equation 19}$$

where $k' = k[OH^-]^y$.

The product of the hydrolysis reaction is $p$-nitrophenol (NP). NP is a yellow compound with an absorbance maximum at 405 nm. More specifically, when a beam of white light with intensity $I_0$ is passed through a sample of NP, light with wavelengths around 405 nm is selectively absorbed, with the remaining light transmitted with an intensity $I$ (Figure 5):

**Figure 5** Schematic showing a cuvette containing a liquid interacting with a beam of light with intensity $I_0$. $I$ is the intensity of the transmitted light; $l$ is the sample path length that the light must travel through.

The absorbance of light by a substance can be related to its concentration using the **Beer-Lambert law** (or **Beer's law**):

$$A = -\log\left(\frac{I}{I_0}\right) = \varepsilon l c \qquad \textbf{Equation 20}$$

A is the absorbance; $I/I_0$ is defined as the transmission (or transmissivity) of the sample; $\varepsilon$ is the molar absorptivity of the substance (otherwise known as the molar extinction coefficient) with units of $M^{-1}cm^{-1}$; $l$ is the path length, or the distance that the light travels through the sample (units of cm); and $c$ is the molar concentration of the sample. The absorbance at 405 nm ($A_{405}$) can therefore be measured and used to determine the concentration of NP at any time during the course of the hydrolysis reaction.

We will use Beer's law in Part II of the procedure to determine the molar extinction coefficient for NP at two different pH values. This is accomplished by measuring the absorbance of NP solution at different NP concentrations. Beer's law is a straight-line equation; by plotting absorbance vs. [NP], a straight line should be generated with a positive slope $= \varepsilon l$. Since NP is formed in a 1:1 stoichiometric ratio from NPA, we can determine the [NPA] remaining at any time by subtracting how much has reacted (i.e., the amount of NPA converted to the product NP) from the initial NPA concentration:

$$[NPA]_t = [NPA]_0 - [NP]_t \qquad \textbf{Equation 21}$$

Using the integrated rate laws, we can then determine the order of the reaction with respect to NPA (i.e., the value of $x$ in Equation 19). If a plot of $[NPA]_t$ vs. time yields a straight line, then $x = 0$ (zeroth order); if a plot of $\ln[NPA]_t$ vs. time yields a straight line, then $x = 1$ (first order); and if a plot of $1/[NPA]_t$ vs. time yields a straight line, then $x = 2$ (second order). The slope of the straight line will give the pseudo rate constant $k'$ (Equation 19). By running the reaction at two different hydroxide concentrations, the value of $y$ in Equation 18 (i.e., the order of the reaction with respect to $[OH^-]$) can also be determined:

$$\frac{k'_1 = k\,[OH^-]_1^y}{k'_2 = k\,[OH^-]_2^y} \quad \rightarrow \quad \frac{k'_1}{k'_2} = \frac{k}{k}\left(\frac{[OH^-]_1}{[OH^-]_2}\right)^y \qquad \textbf{Equation 22}$$

Finally, to determine the value of $k$ (the actual rate constant for the reaction), one substitutes $k'_1$ and $[OH^-]_1$ (or $k'_2$ and $[OH^-]_2$) into Equation 19.

## References

1. Bier M. (1955). Lipases. In *Methods in Enzymology* (Colowick SP, Kaplan NO, eds.), Vol. 1, pp. 627–642. Academic Press, New York.
2. Morillas M, Goble ML, Virden R. "The Kinetics of Acylation and Deacylation of Penicillin Acylase from *Escherichia Coli* Atcc11105: Evidence for Lowered Pk(a) Values of Groups near the Catalytic Centre." *Biochem. J.* 1999; 338: 235–239.
3. Valkova N, Lepine F, Labrie L, Dupont M, Beaudet R. "Purification and Characterization of Prba, a New Esterase from *Enterobacter cloacae* Hydrolyzing the Esters of 4-Hydroxybenzoic Acid (Parabens)." *J. Biol. Chem.* 2003; 278: 12779–12785.
4. Henke E, Bornscheuer UT. "Esterases from *Bacillus Subtilis* and *B. Stearothermophilus* Share High Sequence Homology but Differ Substantially in Their Properties." *Appl. Microbiol. Biotechnol.* 2002; 60: 320–326.

## Procedure

### I. Preparation of Required Solutions

### A. *p*-Nitrophenyl acetate stock solution ($1.00 \times 10^{-4}\,M$ NPA in 3.0% (v/v) methanol/water)

(*Note:* The NPA is not very soluble in water; therefore, methanol is used to dissolve the NPA with subsequent dilution to the appropriate concentration with water.)

1. Measure out 18.1 mg of NPA (181.1 g mol$^{-1}$)—record the mass; place into a 100-mL volumetric flask (or 250-mL Erlenmeyer flask).
2. Add 30.0 mL methanol; mix to dissolve.
3. Dilute to 100.0 mL total volume with deionized water; mix well.
4. Take 1.0 mL of this solution and place in separate container; to this 1.0 mL add 9.0 mL deionized water; mix well.

### B. *p*-Nitrophenol stock solution

(*Note:* The methanol is used in the NP solution simply to make the solvent conditions analogous to the NPA solution.)

5. Measure out 13.9 mg of NP (139.11 g mol$^{-1}$)—record the mass; place into a 100-mL volumetric flask (or 250-mL Erlenmeyer flask).
6. Add 30.0 mL methanol; mix to dissolve.
7. Dilute to 100.0 mL total volume with deionized water; mix well.
8. Take 5.0 mL of this solution and place in separate container; add 45.0 mL deionized water; mix well.

### C. Aqueous methanol solution

9. In a 100-mL graduated cylinder, add 3.0 mL methanol; dilute to 100.0 mL with deionized water; mix well.

### D. NaOH solution ($3.16 \times 10^{-4}$ *M*)

10. Dilute 9.48 mL of standardized 0.0010 *M* NaOH solution to 30 mL using deionized water; mix well.

### E. KCl solution (~3 *M*)

11. Measure out 6.7 g of KCl and place into a 50-mL Erlenmeyer flask.
12. Add 30 mL of deionized water; mix well.

## II. Measurement of Molar Extinction Coefficient of *p*-Nitrophenol at Two [Hydroxide]

13. Prepare a series of four NP samples at different NP concentrations for each hydroxide concentration:

**[Hydroxide] = 0.0010 *M***

14. Label four test tubes (1, 2, 3, and 4).
15. In each test tube, make the following mixtures:

| Tube 1: | 4.0 mL 3 *M* KCl |
| | 4.0 mL NP stock |
| | 4.0 mL 0.0010 *M* NaOH |

| Tube 2: | 4.0 mL 3 *M* KCl |
| | 3.0 mL NP stock |
| | 4.0 mL 0.0010 *M* NaOH |
| | 1.0 mL deionized water |

| Tube 3: | 4.0 mL 3 *M* KCl |
| | 2.0 mL NP stock |
| | 4.0 mL 0.0010 *M* NaOH |
| | 2.0 mL deionized water |

| Tube 4: | 4.0 mL 3 *M* KCl |
| | 1.0 mL NP stock |
| | 4.0 mL 0.0010 *M* NaOH |
| | 3.0 mL deionized water |

**[Hydroxide] = 3.16 × 10⁻⁴ *M***

16. Repeat steps 14 and 15 using tubes labeled 5 to 8, and $3.16 \times 10^{-4}$ *M* NaOH.
17. Prepare two blank solutions (one at each [hydroxide]) by mixing the following solutions together:

    3.0 mL of 3 *M* KCl

    3.0 mL 0.0010 *M* NaOH ([hydroxide] = 0.0010 *M*) or $3.16 \times 10^{-4}$ *M* NaOH ([hydroxide] = $3.16 \times 10^{-4}$ *M*)

    3.0 mL aqueous methanol solution
18. Zero the absorbance measuring device of choice using the blank solution (step 17) for [hydroxide] = 0.0010 *M*.
19. Measure the absorbance at 405 nm for samples 1 through 4; record the absorbance. [*Note:* For each measurement, be sure the cuvette is clean, the outside of the cuvette is wiped clean with a piece of lens paper, and avoid bubbles (which will interfere with the measurement).]
20. Repeat steps 18 and 19 to collect absorbance measurements at [hydroxide] = $3.16 \times 10^{-4}$ *M*.

### III. Determine Reaction Order for *p*-Nitrophenyl Acetate at Two Different [Hydroxide]

**[Hydroxide] = 0.0010 *M***

21. Place the [hydroxide] = 0.0010 *M* blank from step 17 into a cuvette and zero the absorbance measuring instrument at 405 nm; then clean and dry the cuvette.
22. To the empty cuvette, use a pipetter or Pasteur pipette to add the following solutions directly into cuvette and mix well:

    1.0 mL 3 *M* KCl

    1.0 mL 0.0010 *M* NaOH

    As you add each additional solution, add directly into the previous liquid in the cuvette to get good mixing (avoiding bubbles, obviously).
23. Clean the outside of the cuvette with a piece of lens paper, and insert the cuvette into the absorbance measuring device.
24. You will next be adding 1.0 mL of the NPA stock solution. (*Note:* This reaction will begin as soon as it is mixed, and is finished in a short period of time! Add the NPA stock directly to the cuvette into the solution to get good mixing, and begin collecting absorbance data immediately.)
25. Record the absorbance at 405 nm every 10 s for a period of 360 s.

**[Hydroxide] = 3.16 × 10⁻⁴ *M***

26. Repeat steps 21 to 25 making the following changes:
    - Use [hydroxide] = $3.16 \times 10^{-4}$ *M* blank to zero the absorbance.
    - The sample will contain 1.0 mL of $3.16 \times 10^{-4}$ *M* NaOH.

## *Data Collection*

### I. Preparation of Required Solutions

1. Mass of NPA: _____

2. Mass of NP: _____

### II. Measurement of Molar Extinction Coefficient of *p*-Nitrophenol at Two [Hydroxide]

3. Absorbance measurements (405 nm) for NP solutions:

| 0.0010 *M* NaOH 3.16 × 10⁻⁴ *M* NaOH | |
|---|---|
| Tube # | $A_{405}$ |
| 1 | |
| 2 | |
| 3 | |
| 4 | |
| 5 | |
| 6 | |
| 7 | |
| 8 | |

## III.  Determine Reaction Order for *p*-Nitrophenyl Acetate at Two Different [Hydroxide]

[Hydroxide] = 0.0010 *M*

4.  Absorbance (405 nm) as a function of time for NPA hydrolysis reaction:

| Time (s) | $A_{405}$ | Time (s) | $A_{405}$ |
|---|---|---|---|
| 10 | | 190 | |
| 20 | | 200 | |
| 30 | | 210 | |
| 40 | | 220 | |
| 50 | | 230 | |
| 60 | | 240 | |
| 70 | | 250 | |
| 80 | | 260 | |
| 90 | | 270 | |
| 100 | | 280 | |
| 110 | | 290 | |
| 120 | | 300 | |
| 130 | | 310 | |
| 140 | | 320 | |
| 150 | | 330 | |
| 160 | | 340 | |
| 170 | | 350 | |
| 180 | | 360 | |

**[Hydroxide] = $3.16 \times 10^{-4} M$**

5. Absorbance (405 nm) as a function of time for NPA hydrolysis reaction:

| Time (s) | $A_{405}$ | | Time (s) | $A_{405}$ |
|---|---|---|---|---|
| 10 | | | 190 | |
| 20 | | | 200 | |
| 30 | | | 210 | |
| 40 | | | 220 | |
| 50 | | | 230 | |
| 60 | | | 240 | |
| 70 | | | 250 | |
| 80 | | | 260 | |
| 90 | | | 270 | |
| 100 | | | 280 | |
| 110 | | | 290 | |
| 120 | | | 300 | |
| 130 | | | 310 | |
| 140 | | | 320 | |
| 150 | | | 330 | |
| 160 | | | 340 | |
| 170 | | | 350 | |
| 180 | | | 360 | |

## Analysis

### I.  Preparation of Required Solutions

6. Calculate the [NPA] in NPA stock solution (don't forget dilution step):

7. Calculate the [NP] in NP stock solution (don't forget dilution step):

## II. Measurement of Molar Extinction Coefficient of *p*-Nitrophenol at Two [Hydroxide]

**8.** Calculate the [NP] in each test tube:

Tube 1:

Tube 2:

Tube 3:

Tube 4:

Tube 5:

Tube 6:

.

Tube 7:

Tube 8:

**9.** Plot $A_{405}$ vs. [NP] for both hydroxide concentrations:

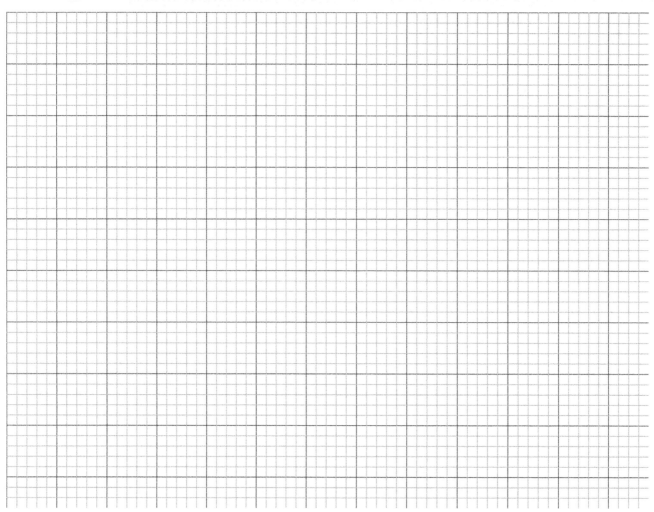

**10.** Using Excel, a graphing calculator, or the least-squares method, fit your data above to a straight line of the form:

$$A_{405} = (\varepsilon \times l) \times [NP] + (y - \text{intercept})$$

| | [OH⁻] = 0.0010 M | [OH⁻] = 3.16 × 10⁻⁴ M |

$[OH^-] = 0.0010\ M \qquad [OH^-] = 3.16 \times 10^{-4}\ M$

(a) Slope of your line (including units): _____ _____

(b) *Y*-intercept of your line (including units): _____ _____

**11.** Determine the molar extinction coefficient of NP at both [hydroxide]:

$[OH^-] = 0.0010\ M \qquad\qquad \varepsilon_{405} = $ _____

$[OH^-] = 3.16 \times 10^{-4}\ M \qquad \varepsilon_{405} = $ _____

## III. Determine Reaction Order for *p*-Nitrophenyl Acetate at Two Different [Hydroxide]

**12.** Calculate the initial [NPA] in each reaction:

**13.** For each [hydroxide], determine the [NP] at each time point using Equation 20:

**[Hydroxide] = 0.0010 *M***

| Time (s) | [NP] | | Time (s) | [NP] |
|----------|------|--|----------|------|
| 10 | | | 190 | |
| 20 | | | 200 | |
| 30 | | | 210 | |
| 40 | | | 220 | |
| 50 | | | 230 | |
| 60 | | | 240 | |
| 70 | | | 250 | |
| 80 | | | 260 | |
| 90 | | | 270 | |
| 100 | | | 280 | |
| 110 | | | 290 | |
| 120 | | | 300 | |
| 130 | | | 310 | |
| 140 | | | 320 | |
| 150 | | | 330 | |
| 160 | | | 340 | |
| 170 | | | 350 | |
| 180 | | | 360 | |

**[Hydroxide] = 3.16 × 10⁻⁴ *M*** 

| Time (s) | [NP] |
|---|---|
| 10 | |
| 20 | |
| 30 | |
| 40 | |
| 50 | |
| 60 | |
| 70 | |
| 80 | |
| 90 | |
| 100 | |
| 110 | |
| 120 | |
| 130 | |
| 140 | |
| 150 | |
| 160 | |
| 170 | |
| 180 | |

| Time (s) | [NP] |
|---|---|
| 190 | |
| 200 | |
| 210 | |
| 220 | |
| 230 | |
| 240 | |
| 250 | |
| 260 | |
| 270 | |
| 280 | |
| 290 | |
| 300 | |
| 310 | |
| 320 | |
| 330 | |
| 340 | |
| 350 | |
| 360 | |

**14.** For each [hydroxide], determine the [NPA] at each time point using Equation 21:

**[Hydroxide] = 0.0010 *M***

| Time (s) | [NPA] |
|---|---|
| 10 | |
| 20 | |
| 30 | |
| 40 | |
| 50 | |
| 60 | |
| 70 | |
| 80 | |
| 90 | |
| 100 | |
| 110 | |
| 120 | |

| Time (s) | [NPA] |
|---|---|
| 130 | |
| 140 | |
| 150 | |
| 160 | |
| 170 | |
| 180 | |
| 190 | |
| 200 | |
| 210 | |
| 220 | |
| 230 | |
| 240 | |

| Time (s) | [NPA] |
|----------|-------|
| 250 | |
| 260 | |
| 270 | |
| 280 | |
| 290 | |
| 300 | |

| Time (s) | [NPA] |
|----------|-------|
| 310 | |
| 320 | |
| 330 | |
| 340 | |
| 350 | |
| 360 | |

[Hydroxide] $= 3.16 \times 10^{-4}\ M$

| Time (s) | [NPA] |
|----------|-------|
| 10 | |
| 20 | |
| 30 | |
| 40 | |
| 50 | |
| 60 | |
| 70 | |
| 80 | |
| 90 | |
| 100 | |
| 110 | |
| 120 | |
| 130 | |
| 140 | |
| 150 | |
| 160 | |
| 170 | |
| 180 | |

| Time (s) | [NPA] |
|----------|-------|
| 190 | |
| 200 | |
| 210 | |
| 220 | |
| 230 | |
| 240 | |
| 250 | |
| 260 | |
| 270 | |
| 280 | |
| 290 | |
| 300 | |
| 310 | |
| 320 | |
| 330 | |
| 340 | |
| 350 | |
| 360 | |

15. You will now use the data for the [hydroxide] $= 0.0010\ M$ reaction to graphically determine the order of the reaction with respect to [NPA]. First, convert [NPA] (#14) at each time point to ln[NPA] and 1/[NPA]:

| Time (s) | ln[NPA] | 1/[NPA] |
|----------|---------|---------|
| 10 | | |
| 20 | | |
| 30 | | |
| 40 | | |
| 50 | | |

| Time (s) | ln[NPA] | 1/[NPA] |
|----------|---------|---------|
| 60 | | |
| 70 | | |
| 80 | | |
| 90 | | |
| 100 | | |
| 110 | | |
| 120 | | |
| 130 | | |
| 140 | | |
| 150 | | |
| 160 | | |
| 170 | | |
| 180 | | |
| 190 | | |
| 200 | | |
| 210 | | |
| 220 | | |
| 230 | | |
| 240 | | |
| 250 | | |
| 260 | | |
| 270 | | |
| 280 | | |
| 290 | | |
| 300 | | |
| 310 | | |
| 320 | | |
| 330 | | |
| 340 | | |
| 350 | | |
| 360 | | |

**16.** Now, plot [NPA] vs. time, ln[NPA] vs. time, and 1/[NPA] vs. time on separate graphs:

[NPA] vs. time:

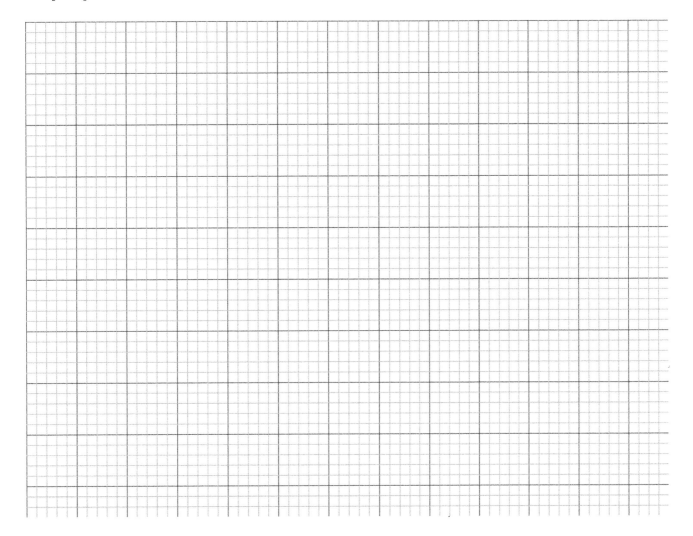

ln[NPA] vs. time:

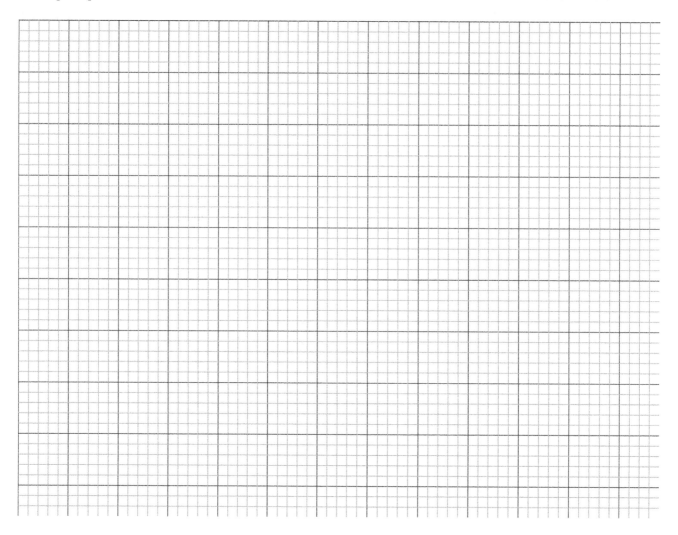

1/[NPA] vs. time:

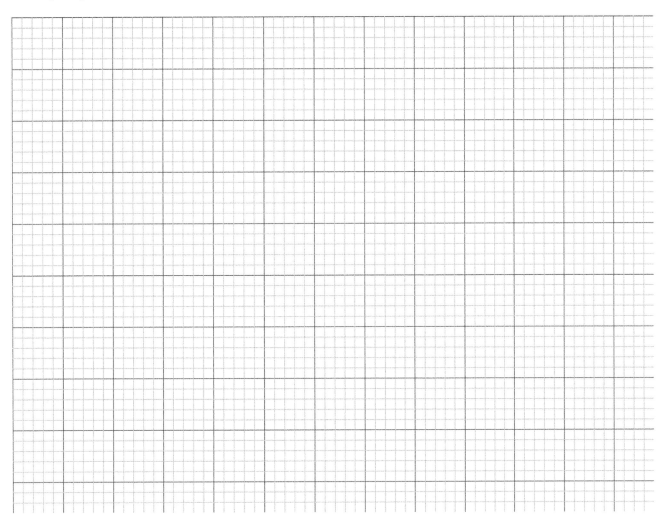

17. Select the plot that is a straight line, and using Excel, a graphing calculator, or the least-squares method, fit your data to a straight line of the form:

$$y = m \times (\text{time}) + (y - \text{intercept})$$

Slope of your line (including units):  _____

Order of the reaction with respect to [NPA]:  _____

18. The slope of the straight line is a pseudo rate constant $k'$ (pseudo because it is equal to the actual rate constant times $[OH^-]^y$). To get the order of the reaction with respect to $[OH^-]$, one needs to have $k'$ at two different $[OH^-]$. Thus, for the $[OH^-] = 3.16 \times 10^{-4}\,M$ data you need to plot either [NPA], ln[NPA], or 1/[NPA] vs. time (pick the one that gave a straight line in #16), fit the data to a straight line, and get the slope:

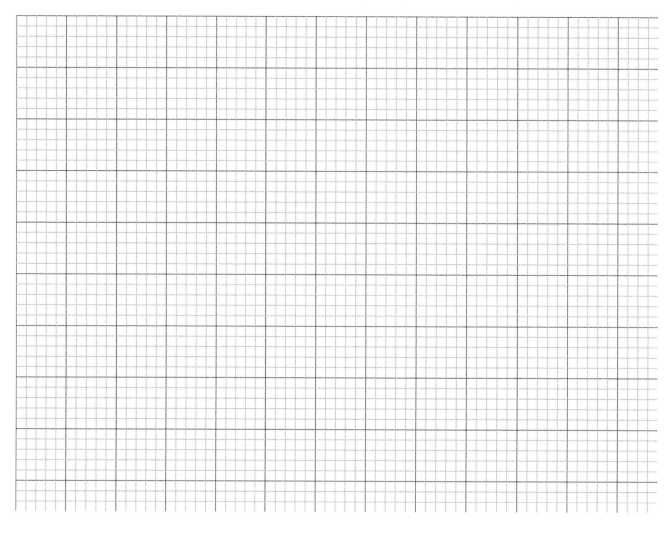

Slope of your line (including units): _____

19. Use Equation 22 to determine the order of the reaction with respect to $[OH^-]$:

Order of the reaction with respect to $[OH^-]$: _____

20. Finally, determine the actual rate constant $k$ for the hydrolysis reaction by using the following equation, as well as one $k'$ and $[OH^-]$ set and the order for $[OH^-]$ determined in #19:

$$k' = k[OH^-]^y$$

Rate constant $k$: _____

**21.** Write the full rate expression (incorporate numerical values for $k$, $x$, and $y$):

$$\text{Rate} = k[\text{NPA}]^x[\text{OH}^-]^y$$

$$= \underline{\hspace{10cm}}$$

## Reflection Questions

1. If you had incorrectly recorded the mass of NPA used to make your stock NPA solution (i.e., had written down a value that was larger than the correct mass), how would that impact the following:
   - $[\text{NPA}]_0$ (#12 in the Analysis section—the initial concentration of NPA in your reaction):

   - [NPA] remaining during the hydrolysis reaction (i.e., #14 in the Analysis section):

2. In the Analysis section, you were asked to use the 0.0010 $M$ NaOH data to graphically determine the order of the reaction based on [NPA]. Why would this be better data to use to graphically determine the order than the $3.16 \times 10^{-4}\ M$ data? (Considered in a different way—how big a difference do you see in the curved vs. linear graphs for the 0.0010 $M$ NaOH data? In the $3.16 \times 10^{-4}\ M$ data? Based on this, why would we select the 0.0010 $M$ NaOH data to graphically determine the exponent for $[\text{OH}^-]$ in the experimental rate law?)

3. To determine the order of the reaction with respect to $[\text{OH}^-]$, you essentially used the initial rates method to get the exponent (i.e., you looked at how changing $[\text{OH}^-]$ affected the rate). Describe how you might determine this order using integrated rate laws.

4. If you had used a 5-mm cuvette instead of a 1-cm cuvette, how would that have impacted the extinction coefficient values calculated in #11 of the Analysis section?

5. Why is the reaction rate dependent on the temperature at which the reaction takes place? What would change about your rate law equation if the temperature was doubled?

6. Based on your knowledge of the hydrolysis reaction, what would happen to the measured rate of the reaction if we removed the OH⁻ from the reaction mixture?

7. Esterases, enzymes that catalyze the hydrolysis of esters in the body, typically contain a metal cation (e.g., $Zn^{2+}$) at their active site. Knowing that the rate of hydrolysis is dependent on [OH⁻], how might the metal cation function to increase the [OH⁻] and, therefore, the rate of the reaction?

## *Connection*

Based on your experience in this lab, draw a connection to something in your everyday life or the world around you (something not mentioned in the background section):

# 18

# Determination of the Equilibrium Constant of Phenolphthalein Dissociation

## Objectives

- Determine the equilibrium constant $K_c$ for a chemical equilibrium.

## Materials Needed

- Deionized water
- $2.05 \times 10^{-4}\ M$ phenolphthalein solution (prepared by instructor)
- Unknown phenolphthalein sample with pH between 8.6 and 9.6 (prepared by instructor)
- pH 8.6 buffer (prepared by instructor)
- pH 8.8 buffer (prepared by instructor)
- pH 9.0 buffer (prepared by instructor)
- pH 9.2 buffer (prepared by instructor)
- pH 9.4 buffer (prepared by instructor)
- pH 9.6 buffer (prepared by instructor)
- 25-mL beakers, Erlenmeyer flasks, scintillation vials, or test tubes (15)
- 10-mL graduated cylinder
- Method for accurately measuring the absorbance of samples at 550 nm. Possible options include:
  - Vernier LabPro handheld device with a SpectroVis Plus spectrophotometer attachment. The associated software LoggerPro must be installed on a computer to allow downloading and manipulation of data. Information about the Vernier device and associated peripherals and software can be obtained from the website http://www.vernier.com (last accessed May 2011).
  - UV/Vis spectrophotometer
  - Spectronic 20 spectrophotometer
- Cuvette compatible with the device used to measure absorbance
- Method for accurately measuring the pH of a solution
- pH 7.0 and 10.0 standards for calibration of pH electrode

## Safety Precautions

- Gloves and safety goggles/glasses should be worn during this lab.
- Phenolphthalein is toxic if ingested, causing damage to the gastrointestinal system.
- Dispose of waste solutions in designated waste containers as indicated by your instructor.

## Background

### EQUILIBRIUM

Many physical processes and chemical reactions establish a state of equilibrium under appropriate conditions. When at equilibrium, the rates of the forward and reverse reactions of a reversible process are equivalent, yielding no net change in the amounts of products and reactants. The system will appear to have stopped changing, although this is not true; an equilibrium is dynamic—both forward and reverse reactions are still occurring, just at equivalent rates.

As an example, let's consider the simple equilibrium:

$$A(g) \rightleftharpoons 2B(g) \qquad \qquad \textbf{Reaction 1}$$

The rate laws for the forward and reverse reactions can be written as:

$$\text{rate}_{\text{forward}} = k_f\, [A]^x \qquad \qquad \textbf{Equation 1a}$$

$$\text{rate}_{\text{reverse}} = k_r\, [B]^y \qquad \qquad \textbf{Equation 1b}$$

If the reaction container initially contains pure A (i.e., $[B]_{\text{init}} = 0$), then $\text{rate}_{\text{forward}}$ will necessarily be larger than $\text{rate}_{\text{reverse}}$, which will equal zero. As the reaction proceeds forward, A is converted to B, so [A] decreases and [B] increases; this leads to a decrease in $\text{rate}_{\text{forward}}$ and an increase in $\text{rate}_{\text{reverse}}$. Once the system reaches equilibrium, [A] and [B] are no longer changing, yielding constant and equivalent values for $\text{rate}_{\text{forward}}$ and $\text{rate}_{\text{reverse}}$. A plot of reactant and product concentrations vs. time clearly shows how the concentrations become constant at some time during the reaction—it is at this point that equilibrium is reached (Figure 1a). Alternatively, in Figure 1b, the reaction rate is plotted vs. time, and it shows that $\text{rate}_{\text{forward}}$ and $\text{rate}_{\text{reverse}}$ become equal at the same time that the concentrations become constant—this again is when equilibrium is established for the reaction.

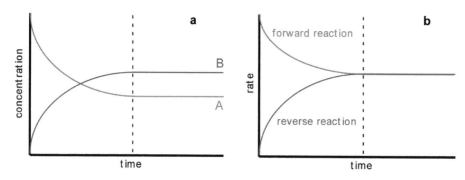

**Figure 1** Concentration vs. time (a) and reaction rate vs. time (b) plots for Reaction 1, where the initial conditions contain pure A, and $[B]_{\text{init}} = 0$. The dotted line in each plot indicates the point in time when the concentrations become constant (a), and when the forward and reaction rates become equivalent (b).

A separate experiment in which the reaction container initially contains pure B (i.e., $[A]_{\text{init}} = 0$) will show a similar outcome. Under these conditions the initial $\text{rate}_{\text{reverse}}$ will be significantly larger than $\text{rate}_{\text{forward}}$, which will be equal to zero. As the reaction proceeds in the reverse direction, B is converted to A, so [B] decreases and [A] increases; this leads to an increase in $\text{rate}_{\text{forward}}$ and a decrease in $\text{rate}_{\text{reverse}}$. Once the system reaches equilibrium, [A] and [B] again are no longer changing, yielding constant values for $\text{rate}_{\text{forward}}$ and $\text{rate}_{\text{reverse}}$. In addition, $\text{rate}_{\text{forward}}$ will again equal $\text{rate}_{\text{reverse}}$ at equilibrium. For this experiment, the concentration vs. time and rate vs. time plots in Figure 2 have similar features to those in Figure 1—namely, at equilibrium the concentrations are constant and $\text{rate}_{\text{forward}}$ and $\text{rate}_{\text{reverse}}$ are equivalent.

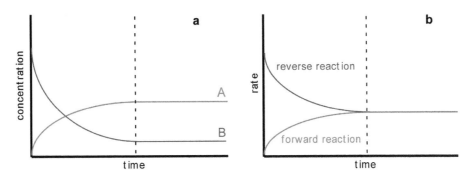

**Figure 2** Concentration vs. time (a) and reaction rate vs. time (b) plots for Reaction 1, where the initial conditions contain pure B, and $[A]_{init} = 0$. The dotted line in each plot indicates the point in time when the concentrations become constant (a), and when the forward and reaction rates become equivalent (b).

Several important points are made in these two experiments: (1) equilibrium can be established from more than one set of initial reactant/product concentrations; (2) at equilibrium, concentrations of species in the reaction are constant; and (3) at equilibrium, the forward and reverse reaction rates are equal.

## EQUILIBRIUM CONSTANTS AND REACTION QUOTIENTS

For the following reaction:

$$H_2(g) + I_2(g) \rightleftharpoons 2HI(g)$$ **Reaction 2**

a series of experiments was performed with different starting concentrations, all at 25°C. After equilibrium is reached for each reaction, the equilibrium concentrations were determined with results listed in Table 1.

| Table 1 | Initial and final concentrations of all reactants and products in Reaction 2. | | | | | | |
| --- | --- | --- | --- | --- | --- | --- | --- |
| Experiment | $[H_2]_{init}$ (*M*) | $[I_2]_{init}$ (*M*) | $[HI]_{init}$ (*M*) | $[H_2]_{equil}$ (*M*) | $[I_2]_{equil}$ (*M*) | $[HI]_{equil}$ (*M*) | $\dfrac{[HI]^2_{equil}}{[H_2]_{equil}[I_2]_{equil}}$ |
| 1 | 0.500 | 1.00 | 0 | $2.82 \times 10^{-3}$ | 0.504 | 0.994 | 697 |
| 2 | 0 | 0 | 0.525 | 0.0183 | 0.0183 | 0.488 | 711 |
| 3 | 1.56 | 0.232 | 0 | 1.33 | $2.28 \times 10^{-4}$ | 0.464 | 710 |

As we can see, the equilibrium concentrations are very different from experiment to experiment. What *is* similar for all experiments, however, is the quotient in the far right column (the minor variations in the quotient values are due to experimental variations). This quotient is composed of the product concentrations divided by the reactant concentrations with each concentration raised to a power equivalent to the corresponding coefficient from the *balanced* chemical equation (see Reaction 2). This quantity is called the **reaction quotient,** or $Q_c$ (where the "c" subscript indicates the expression is written in terms of concentrations). For any general equilibrium, the reaction quotient can be written as

$$a\,A + b\,B \rightleftharpoons c\,C + d\,D \quad \text{reaction quotent} = Q_c = \frac{[C]^c[D]^d}{[A]^a[B]^b}$$ **Equation 2**

When the system is at equilibrium, equilibrium concentrations can be substituted into Equation 2, and the reaction quotient (which is constant at equilibrium) is named the **equilibrium constant,** or $K_c$.

The equilibrium constant expression shown in Equation 2 can also be derived from the forward and reverse reaction rates, which are equivalent at equilibrium.

$$\text{rate}_{\text{forward}} = \text{rate}_{\text{reverse}}$$  **Equation 3**

Substituting the rate law expressions into Equation 3 yields Equation 4:

$$k_f[A]^a[B]^b = k_r[C]^c[D]^d$$  **Equation 4**

Rearrangement yields the equilibrium constant expression from Equation 2:

$$\frac{k_f}{k_r} = K_c = \frac{[C]^c[D]^d}{[A]^a[B]^b}$$  **Equation 5**

There are several important points regarding the reaction quotient and the equilibrium constant:

1. $K_c$ and $Q_c$ are written using concentrations.
2. The exponents come from the coefficients of the **balanced** chemical equation.
3. $Q_c$ and $K_c$ have the same equation—they differ in that $Q_c$ can be calculated using **any** set of concentrations, while $K_c$ is calculated using equilibrium concentrations only.
4. $Q_c$ and $K_c$ are typically reported without units (you can read or ask your instructor about **activities**).
5. $Q_c$ and $K_c$ are temperature dependent.
6. $Q_c$ will change as a reaction proceeds toward equilibrium, with the value approaching that of $K_c$. In fact, the values of $Q_c$ and $K_c$ can be compared to determine in which direction the reaction must proceed (i.e., forward toward products, or in reverse toward reactants); when $Q_c < K_c$, the reaction will proceed in the forward direction ($Q_c$ increases in this direction) until $Q_c = K_c$ (system at equilibrium); when $Q_c > K_c$, the reaction will proceed in reverse ($Q_c$ decreases in this direction) until $Q_c = K_c$; when $Q_c = K_c$, the system is at equilibrium, and the reaction will not proceed in either direction.
7. The use of rate equations to derive the equilibrium constant expression (Equations 3 to 5) is valid for simple one step reactions; for multistep reactions, the derivation is more involved, but the resulting equilibrium constant expression (Equation 5) will still have the same form.

When writing $Q_c$ and $K_c$, one must take into consideration the **phase** of each compound. Since $Q_c$ and $K_c$ are defined in terms of concentrations, each species in Equations 2 and 5 must have a concentration associated with it that can change during the course of the reaction. Substances in solution (aq) obviously qualify and so are included. Gaseous species (g) also have concentrations that can change and so are included (the concentration of a gas depends on the volume of the container in which it is contained). Pure solids and liquids are omitted from the equilibrium constant and reaction quotient expressions, since their ratio of moles (mass) to volume is constant at a constant $T$ (technically, these "constants" are included in $K_c$ and $Q_c$—the equation contains only those species that are considered variable). So for the heterogeneous equilibria listed below (heterogeneous indicates more than one phase), the corresponding $K_c$ expressions are written:

$$CaCO_3(s) \rightleftharpoons CaO(s) + CO_2(g) \qquad\qquad K_c = [CO_2]$$

$$Cu(s) + 2Ag^+(aq) \rightleftharpoons Cu^{2+}(aq) + 2Ag(s) \qquad K_c = \frac{[Cu^{2+}]}{[Ag^+]^2}$$

Another variation for writing equilibrium constants involves reactions containing gaseous species. In this case, $K$ can be written in terms of either concentrations (giving a $K_c$) or partial pressures (in atm) of the individual gases (giving a $K_P$). As an example, for the chemical reaction between nitrogen and hydrogen to yield ammonia, the following $K$ expressions can be written:

$$N_2(g) + 3H_2(g) \rightleftharpoons 2NH_3(g) \qquad K_c = \frac{[NH_3]^2}{[N_2][H_2]^3} \quad \text{and} \quad K_P = \frac{P_{NH_3}^2}{P_{N_2}P_{H_2}^3}$$

The relationship between $K_P$ and $K_c$ for a given reaction is based on the ideal gas law:

$$P_{gas} = \frac{n}{V}RT = [gas]RT \qquad \qquad \textbf{Equation 6}$$

Substituting Equation 6 into $K_P$ yields Equation 7, which allows a direct conversion between $K_P$ and $K_c$:

$$K_P = K_c\,(RT)^{\Delta n_{gas}} \qquad \qquad \textbf{Equation 7}$$

where $\Delta n_{gas}$ is the change in moles of gas in the balanced chemical reaction [(moles of gaseous products)–(moles of gaseous reactants)], and $R$ = ideal gas constant = 0.08206 L atm mol$^{-1}$ K$^{-1}$.

## EQUILIBRIUM CALCULATIONS

The equilibrium constant expression can be used to perform two basic types of calculations: (1) given a set of equilibrium concentrations, one can calculate the $K_c$ for a reaction; and (2) given a set of initial reactant/product concentrations and a numerical value for $K_c$, one can determine the equilibrium concentrations of all reactants and products. The first type is relatively straightforward. In this section, the focus will be on the second variety.

The approach used for calculating equilibrium concentrations when given $K_c$ and an initial set of concentrations involves the construction of an equilibrium table (or **ICE** table, named after the rows of the table). The steps are outlined next; we defer to your general chemistry textbook to provide detailed examples.

1. Construct an **ICE** table; fill in the initial concentrations (including any that are zero) in the "Initial (I)" line of the table.
2. Use initial concentrations to calculate $Q_c$; compare $Q_c$ and $K_c$ to determine the direction in which the reaction will proceed to reach equilibrium. (*Note:* No species will have a zero concentration at equilibrium, so the reaction will always proceed in the direction that increases the concentration of a reactant or produce with [species]$_{initial}$ = 0.)
3. Define "$x$" as the amount of a particular species consumed—this is entered in the "Change (C)" line of the table; use the stoichiometry of the reaction to define the amount of other species consumed or produced in terms of $x$.
4. For each species in the "Equilibrium (E)" line of the table, add the change in concentration to the initial concentration to get the equilibrium concentration.
5. Substitute the equilibrium concentrations into the $K_c$ expression and solve for $x$. At this point, there are several mathematical strategies that can be used, including use of the quadratic equation and an approximation (or successive approximations) method.
6. Using the calculated value of $x$, determine the concentrations of all other species at equilibrium.
7. Check your work by plugging the calculated equilibrium concentrations into the equilibrium constant expression. The result should be very close to the given $K_c$.

## THIS LAB

The goals of this lab are to (1) determine $K_c$ for the acid-base indicator phenolphthalein and (2) determine the pH of an unknown phenolphthalein solution based on its absorbance. Phenolphthalein ($C_{20}H_{14}O_4$; molar mass = 318.32 g mol$^{-1}$) is a weak acid that reacts with water and dissociates into a weakly basic anion and a hydronium ion; it is considered a weak acid in that this dissociation does not go to completion—that is, it establishes an equilibrium:

$$HIn + H_2O \rightleftharpoons In^- + H_3O^+ \qquad K_c = \frac{[In^-][H_3O^+]}{[HIn]} \qquad \textbf{Reaction 3 and Equation 8}$$

You might recognize phenolphthalein as an acid/base indicator that was used during an earlier acid-base titration exercise. As an acid-base indicator, phenolphthalein is used to identify the endpoint of acid-base reactions, where it changes from colorless to a light pink color. The endpoint of a titration is defined as the point where the stoichiometric equivalence point is reached, as defined by a balanced chemical equation.

Phenolphthalein changes color when its protonation state changes: the HIn form ("acidic" form) is colorless in solutions with relatively high $[H_3O^+]$/low $[OH^-]$ (i.e., acidic solutions); the In$^-$ form ("basic" form) is a pink/fuchsia color in solutions with relatively low $[H_3O^+]$/high $[OH^-]$ (i.e., basic solutions). The relative position of the equilibrium (i.e., relative concentrations of HIn and In$^-$) depend on $[H_3O^+]$. The two relevant forms of phenolphthalein are shown in Figure 3:

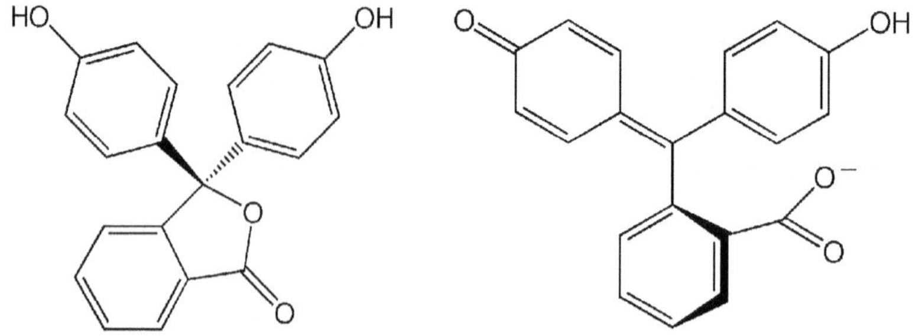

HIn: pH 0 to 8.2          In$^-$: pH 8.2 to 10

Figure 3   Acidic (left) and basic (right) forms of phenolphthalein.

The loss of a proton from the acidic form leads to a change in the hybridization of the central carbon from $sp^3$ to $sp^2$, allowing the two aromatic rings to become approximately coplanar (increases conjugation). This structural rearrangement changes the energy of the photons that are absorbed by the molecule—HIn absorbs photons of light outside the visible range (ultraviolet region) and so is colorless, while In$^-$ absorbs photons within the visible range, and so it is colored (absorbs in the blue-green region, so appears pink/red—absorbs at 550 nm).

Besides being a popular acid-base indicator, phenolphthalein has other uses:

- It was previously used as an active ingredient in laxatives such as Ex-Lax.
- It is used in a blood detection test known as the Kastle-Meyer test (think *CSI!*). To a given area at a crime scene or sample, HIn (dissolved in alcohol) is added, followed by hydrogen peroxide ($H_2O_2$); if the sample turns pink, then the presence of blood is confirmed (although a positive will occur for blood from any mammal).

In this lab, you will prepare a series of phenolphthalein samples at different $[H_3O^+]$ values. Each solution should contain different relative concentrations of HIn and In$^-$ in equilibrium, and should therefore have different color intensities. Based on Equation 8, we can determine an experimental value for $K_c$ by measuring the values of [HIn], [In$^-$] and $[H_3O^+]$. The value of $[H_3O^+]$ is measured experimentally as the pH of the solution, where pH is defined as $-\log[H_3O^+]$. The pH in each solution will be fixed at a constant value using a glycine buffer (a buffer is a substance that limits changes in pH by reacting with/neutralizing added acid or base). To determine [HIn] and [In$^-$] at different $[H_3O^+]$ values, you will utilize an equilibrium (ICE) table:

$$HIn + H_2O \rightleftharpoons In^- + H_3O^+$$

| | | | | |
|---|---|---|---|---|
| I [initial] | | | | |
| C (change) | | | | |
| E [equilibrium] | | | | |

**Initial line in table:**

[HIn]$_{init}$ is known

[In$^-$]$_{init}$ = 0

[H$_3$O$^+$]$_{init}$ is known (pH of the glycine buffer)

**Equilibrium line in table:**

$[In^-]_{equil}$ will be determined by measuring $A_{550}$ (the absorbance at 550 nm)

$[H_3O^+]_{equil} = [H_3O^+]_{init}$ (the buffer keeps this at a constant value; even though $H_3O^+$ is generated during the conversion of HIn to $In^-$, the buffer is designed to keep a constant $[H_3O^+]$; therefore, regardless of the $H_3O^+$ generated by the equilibrium, the $[H_3O^+]_{equil}$ is dictated by the buffer, so $[H_3O^+]_{equil} = [H_3O^+]_{init}$)

$[HIn]_{equil} = [HIn]_{init} - [In^-]_{equil}$

Once the equilibrium values are determined, one can substitute $[HIn]_{equil}$, $[In^-]_{equil}$, and $[H_3O^+]_{equil}$ into Equation 8 to calculate $K_c$ for that specific $[H_3O^+]$. Then, by varying $[H_3O^+]$, you will determine a series of $K_c$ values, which can be averaged together to get a final average $K_c$ value.

There is also a graphical method that works well to calculate $K_c$. If one takes the $-\log$ of both sides of Equation 8 and rearranges, the result is Equation 9:

$$-\log K_c = -\log[H_3O^+] - \log\frac{[In^-]}{[HIn]}$$

$$pH = \log\frac{[In^-]}{[HIn]} - \log K_c \qquad \textbf{Equation 9}$$

Equation 9 is the equation of a straight line ($y = mx + b$). Thus, a plot of pH vs. $\log\dfrac{[In^-]}{[HIn]}$ should yield a straight line with a slope $= 1$ and a $y$-intercept equal to $-\log K_c$.

# References

1. Khanolkar MM, Sirsat AV, Walvekar SS, Bhansali MS, Desouza LJ. "Modified Method for Determination of Serum Beta-Glucuronidase: A Comparative Study Using P-Nitrophenyl Glucuronide and Phenolphthalein Glucuronide as Substrate in Gastrointestinal Tract Carcinomas." *Indian Journal of Clinical Biochemistry.* 1997; 12: 67–70.

# Procedure

## I. Preparation of Required Solutions

### Phenolphthalein Solutions

1. In clean containers, each student/group should obtain the following solutions from the instructor:
   - ~8 mL of the phenolphthalein solution
   - ~20 mL of each of six buffer solutions, labeled pH 8.8 through pH 9.6
   - ~10 mL of an unknown solution

2. Label each of six clean glass containers (25-mL beakers or flasks, scintillation vials, or test tubes) with one of the following labels: 8.6, 8.8, 9.0, 9.2, 9.4, and 9.6 (these will be the approximate pH values for each of the samples you will be generating).

3. In each container, add 1.0 mL of the phenolphthalein solution and 9.0 mL of the appropriate buffer solution; mix well.

### Blank Solution

4. Mix 1.0 mL of deionized water with 9.0 mL of the pH 9.2 buffer solution; mix well. [*Note:* One blank solution will work for all samples because the buffer solution being used (based on glycine) does not absorb significantly around 550 nm, where you will be collecting absorbance data, and this is not affected by the solution pH.]

## Unknown Solution

5. Label a clean container with "unknown."
6. Add 1.0 mL of the phenolphthalein solution and 9.0 mL of the unknown solution; mix well.

### II. Measurement of Absorbance at 550 nm of Phenolphthalein Solutions

7. Zero the absorbance measuring device of choice using the blank solution (step 4).
8. Measure the absorbance at 550 nm for the six phenolphthalein samples (samples 8.8 through 9.6); record the absorbance for each. [*Note:* For each measurement, be sure the cuvette is clean, the outside of the cuvette is wiped clean with a piece of lens paper, and avoid bubbles (which will interfere with the measurement).]
9. Measure the absorbance at 550 nm for the unknown sample; record the absorbance.

### III. Measurement of pH of Phenolphthalein Solutions

10. Calibrate the pH electrode using the appropriate pH standards (you will want to be sure the electrode is calibrated for the pH range 7 through 10) and the pH measuring device of choice.
11. Measure the pH of the six phenolphthalein samples (samples 8.8 through 9.6); record the pH for each.

## Data Collection

### II. Measurement of Absorbance at 550 nm of Phenolphthalein Solutions

1. Record absorbance at 550 nm for the phenolphthalein samples:

| Sample | $A_{550}$ |
|--------|-----------|
| 8.6 | _____ |
| 8.8 | _____ |
| 9.0 | _____ |
| 9.2 | _____ |
| 9.4 | _____ |
| 9.6 | _____ |

2. Record absorbance at 550 nm for the unknown sample:

| Sample | $A_{550}$ |
|--------|-----------|
| Unknown | _____ |

### III. Measurement of pH of Phenolphthalein Solutions

3. Record the pH of the phenolphthalein samples:

| Sample | Measured pH |
|--------|-------------|
| 8.6 | _____ |
| 8.8 | _____ |
| 9.0 | _____ |
| 9.2 | _____ |
| 9.4 | _____ |
| 9.6 | _____ |

**4.** Calculate $[H_3O^+]_{init}$ from measured pH for each phenolphthalein sample:

| Sample | $[H_3O^+]_{init}$ |
|---|---|
| 8.6 | _____ |
| 8.8 | _____ |
| 9.0 | _____ |
| 9.2 | _____ |
| 9.4 | _____ |
| 9.6 | _____ |

**5.** Calculate $[In^-]_{equil}$ from measured $A_{550}$ using Beer's law for each phenolphthalein sample [molar extinction coefficient for phenolphthalein ($\lambda = 550$ nm) $= 29{,}300\ M^{-1}\ cm^{-1}$]:[1]

| Sample | $[In^-]_{equil}$ |
|---|---|
| 8.6 | _____ |
| 8.8 | _____ |
| 9.0 | _____ |
| 9.2 | _____ |
| 9.4 | _____ |
| 9.6 | _____ |

**6.** Calculate $[In^-]_{equil}$ from measured $A_{550}$ using Beer's law for unknown sample [molar extinction coefficient for phenolphthalein ($\lambda = 550$ nm) $= 29{,}300\ M^{-1}\ cm^{-1}$]:[1]

| Sample | $[In^-]_{equil}$ |
|---|---|
| sample | _____ |

**7.** Calculate $[HIn]_{equil}$ for each phenolphthalein sample:

| Sample | $[HIn]_{equil}$ |
|---|---|
| 8.6 | _____ |
| 8.8 | _____ |
| 9.0 | _____ |
| 9.2 | _____ |
| 9.4 | _____ |
| 9.6 | _____ |

**8.** Calculate $[HIn]_{equil}$ for unknown sample:

| Sample | $[HIn]_{equil}$ |
|---|---|
| Unknown | _____ |

## Determine $K_c$ by Using Equilibrium Constant Equation

9. Calculate $K_c$ for each phenolphthalein sample:

| Sample | $K_c$ |
|--------|-------|
| 8.6 | _____ |
| 8.8 | _____ |
| 9.0 | _____ |
| 9.2 | _____ |
| 9.4 | _____ |
| 9.6 | _____ |

10. Calculate the average $K_c$ for phenolphthalein:

average $K_c$ = _____

## Determine $K_c$ by Using Graphical Method

11. Calculate $\log \dfrac{[\text{In}^-]}{[\text{HIn}]}$ for each phenolphthalein sample:

| Sample | $\log \dfrac{[\text{In}^-]}{[\text{HIn}]}$ |
|--------|-------|
| 8.6 | _____ |
| 8.8 | _____ |
| 9.0 | _____ |
| 9.2 | _____ |
| 9.4 | _____ |
| 9.6 | _____ |

**12.** Plot pH vs. $\log \dfrac{[In^-]}{[HIn]}$:

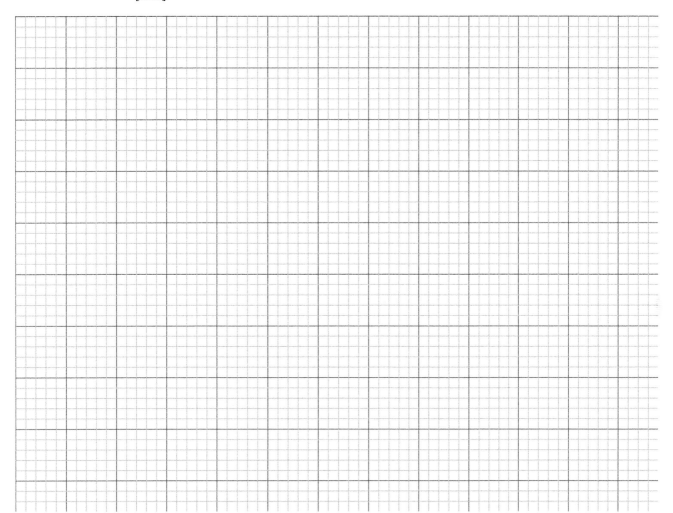

**13.** Using Excel, a graphing calculator, or the least-squares method, fit your data above to a straight line of the form:

$$pH = \log\dfrac{[In^-]}{[HIn]} + (-\log K_c)$$

(a) Slope of your line (including units): _____

(b) y-intercept of your line (including units): _____

**14.** Calculate value of $K_c$ using results in #13:

Value of $K_c$ = _____

## Determine pH of Unknown Sample

**15.** Determine the pH of the unknown sample:

pH of unknown sample = _____

description of the method you used to determine pH:

## Reflection Questions

1. The value for the $K_c$ of phenolphthalein is commonly reported as $5.01 \times 10^{-10}$. Calculate the percent error for the $K_c$ value that you determined two different ways:
   (a) Using the equilibrium constant equation

   (b) Using the graphing method

2. Based on the average value of $K_c$ that you determined, classify the phenolphthalein reaction as either reactant- or product-favored and explain your choice.

3. Is there a relationship between your calculated $K_c$ value and the pH of the sample? If so, why might that be occurring?

4. What value for $K_c$ would you get if you used the predicted pH value for the pH 9.0 buffer and not the measured value?

5. Assume that for the pH 9.2 phenolphthalein sample, you used half as much phenolphthalein solution (i.e., your sample is 9 mL buffer, 0.5 mL phenolphthalein solution, and 0.5 mL water). What should happen to the value of $K_c$ you determine?

6. What are the possible sources of errors in this lab? How might each affect your calculated $K_c$ value?

7. On your straightline plot, should the line pass through the origin? Explain your answer.

## Connection

Based on your experience in this lab, draw a connection to something in your everyday life or the world around you (something not mentioned in the background section):

# 19

# Le Châtelier's Principle: On the Effect of Concentration and Temperature on Equilibrium

## Objectives

- Observe effect of changes in reactant and product concentrations on equilibrium.
- Observe effect of temperature on equilibrium.

## Materials Needed

- Deionized water
- Ice
- Calcium hydroxide [$Ca(OH)_2$] solid
- 6 $M$ Hydrochloric acid (HCl) solution (prepared by instructor)
- 0.1 $M$ HCl solution (prepared by instructor)
- 0.1 $M$ Iron(III) nitrate [$Fe(NO_3)_3$] solution (prepared by instructor)
- 0.1 $M$ Potassium thiocyanate (KSCN) solution (prepared by instructor)
- 0.1 $M$ Silver nitrate ($AgNO_3$) solution (prepared by instructor)
- Cobalt(II) chloride ($CoCl_2$) dissolved in ethanol solution (prepared by instructor)
- 1.0 $M$ Calcium chloride ($CaCl_2$) solution (prepared by instructor)
- Bromophenol blue stock solution (prepared by instructor)
- 0.1 $M$ Sodium hydroxide (NaOH) solution (prepared by instructor)
- 6 $M$ NaOH solution (prepared by instructor)
- 250-mL Erlenmeyer flask (1)
- 250-mL beakers (2)
- Test tubes (13 $\times$ 100 mm, 9-mL volume) (15)
- Test tube rack
- Test tube clamp
- Hot plate
- Ring stand and clamp

## Safety Precautions

- Gloves and safety goggles/glasses should be worn during this lab.
- Methanol is flammable—keep away from open flames.
- Concentrated hydrochloric acid is a strong acid and is toxic by ingestion and inhalation and corrosive to skin and eyes; avoid contact with body tissues. It must be stored and used in a laboratory hood.

- Concentrated sodium hydroxide is a strong base that can cause burns; avoid contact with body tissues. It must be stored and used in a laboratory hood.
- In general, avoid contact with skin or eyes or ingestion of the solid and solution-based reagents in this lab.
- Dispose of waste solutions in designated waste containers as indicated by your instructor.

## *Background*

In Laboratory 18, a chemical equilibrium was studied and its equilibrium constant ($K_c$) determined. The focus of the current lab is to investigate how changes in the reactant or product concentrations of an equilibrium, as well as changes in temperature, cause predictable changes in the position of an equilibrium.

## LE CHÂTELIER'S PRINCIPLE

When a system is at equilibrium, the rates of the forward and reverse reactions are equivalent, yielding no net change in the amounts of products and reactants. Equilibria are dynamic—both forward and reverse reactions are still occurring, just at equivalent rates. Once at equilibrium, anything that is done to the system that affects one of the rates preferentially will disrupt the equilibrium, causing the reaction/process to move in the direction with the greater rate. The French chemist/engineer Henri-Louis Le Châtelier studied chemical equilibria and proposed what has come to be known as **Le Châtelier's principle:**

> when a stress is applied to a system at equilibrium, the system will respond by shifting in the direction that minimizes/counteracts the effect of the stress in order to reestablish equilibrium, where "stress" refers to a change in one or more reactant/product concentrations, pressure or temperature.

A shift to the right (resulting in more products formed) results from a change that increases the forward rate, while a shift to the left (resulting in a decrease in products) results from a change that increases the reverse reaction rate. We can make qualitative predictions about the change that will occur based on what stress is applied. Given next are several factors that affect an equilibrium.

## 1. Change in [Reactant] or [Product]

Changing the concentration of one or more reactants or products can be accomplished either by addition of one of the species to increase its concentration, or by removing some fraction of a reactant or product, often by addition of something that reacts and decreases its concentration. A change in concentration changes the value of the reaction quotient ($Q$). If reactant is added, there is an increase in the forward reaction rate, while removal of product decreases the reverse reaction rate. In both cases, the net flow of the reaction is forward. By moving in the forward direction, reactants are consumed, which counteracts the effect of the initial concentration change. The forward rate increases initially until a set of concentrations are reached where the forward and reverse rates are again equivalent; this is a new equilibrium position. In terms of $Q$ vs. $K$, the initial addition of reactant (or removal of product) causes $Q < K$; as the reaction proceeds forward, product concentrations increase and reactant concentrations decrease, causing $Q$ to increase until $Q = K$. The system is again at equilibrium and no further change is observed. For example, for the following chemical reaction:

$$Cr_2O_7^{2-}(aq) + 2OH^-(aq) \rightleftharpoons 2CrO_4^{2-}(aq) + H_2O(l) \qquad \textbf{Reaction 1}$$

addition of NaOH will shift the equilibrium to the right; addition of $K_2CrO_4$ will shift the equilibrium to the left; and addition of HCl will shift the equilibrium to the left ($H^+$ reacts with $OH^-$ to form water, removing $OH^-$ from the equilibrium).

## 2. Change in Volume (and thus Pressure) In Gaseous Reactions

If an equilibrium contains gases, then changes to the volume of the reaction container can have an impact. The ideal gas law can be used to show this effect:

$$P = \left(\frac{n}{V}\right) RT = \text{(concentration)} \, RT$$

As one can see, volume changes affect both the concentration and the partial pressure of the gas. A decrease in volume of the container increases both the partial pressure of the gas and its concentration. Since the whole reaction is in the same container, a volume change impacts all gases in the reaction (volume changes have little impact on solids and liquids in the equilibrium). For the following equilibrium:

$$3H_2(g) + N_2(g) \rightleftharpoons 2NH_3(g) \qquad \textbf{Reaction 2}$$

a decrease in container volume will increase the partial pressure of every gas in the reaction. Le Châtelier's principle indicates that the equilibrium will shift to counteract the stress—since the total pressure is increased by the volume change, the equilibrium will shift toward the product side, which contains fewer moles of gas in the balanced reaction (thus decreasing the number of moles in the container and, thus, the total pressure). In terms of concentrations, the decrease in container volume increases the concentration of everything. With more total moles on the reactant's side, there is a net increase of reactant concentrations; thus, the reaction will shift right to decrease the reactant concentrations. In the next equilibrium,

$$H_2(g) + I_2(g) \rightleftharpoons 2HI(g) \qquad \textbf{Reaction 3}$$

volume changes will have no impact since the total number of moles of gas on both sides is equal. So volume changes will lead to equivalent concentration (or pressure) changes for both sides of the equilibrium and no shift is observed. Finally, if the reaction contains NO gases, then volume/pressure changes have minimal to no effect on the equilibrium position. Addition of an inert gas to a gaseous equilibrium (inert indicates that the added gas will not participate in a chemical reaction with any species in the equilibrium) has no impact on the equilibrium position because it does not change either the concentration or the partial pressure of any species in the reaction.

## 3. Change the Temperature

To understand the impact that temperature has on the equilibrium, it is necessary to know if the reaction is endothermic or exothermic. As an example, the following equilibrium is exothermic:

$$3H_2(g) + N_2(g) \rightleftharpoons 2NH_3(g) + \text{heat} \qquad \textbf{Reaction 4}$$

The heat change in this equilibrium can be considered as a product (i.e., heat is produced/released by the forward reaction). If the temperature is raised for this reaction, energy is being added—this is effectively an increase in a product of the reaction. Le Châtelier's principle indicates that the equilibrium will therefore shift to the left (toward the reactants) to counteract the increase of energy. In general, when the temperature is increased, the equilibrium will shift in the direction that produces an **endothermic** change, while a decrease in temperature leads to a shift in the direction that produces an **exothermic** change.

Changing the temperature also has an impact on the value of the equilibrium constant $K$—this is in contrast to the other stresses already described, which impact $Q$ but not $K$. When the temperature is changed, the reaction will shift to one side to establish a new equilibrium position (i.e., new reactant and product concentrations). This change was not a result of initially changing the concentration of anything. Therefore, a new value of $K$ is generated. So, for Reaction 4, the increase in temperature shifts the equilibrium to the left, which decreases the product concentrations and increases the reactant concentrations, resulting in a decrease in the value of $K$.

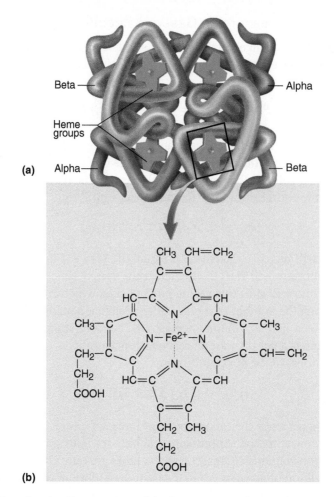

**Figure 1** (a) Ribbon illustration showing the structure of the homotetrameric protein hemoglobin, the oxygen-transporting protein present in vertebrate animals. Cylinders approximate the location of the protein backbone—two monomers are colored blue and two are colored brown. The heme groups, which are used to bind the $O_2$, are highlighted in red (one per monomer, with the $Fe^{2+}$ in each heme represented by a blue sphere). (b) Structure of the heme group showing the coordination of the $Fe^{2+}$ by four heme nitrogen atoms; a fifth coordination site of the $Fe^{2+}$ is occupied by the nitrogen from a histidine amino acid side chain (not shown); this leaves a sixth position in the $Fe^{2+}$ coordination geometry open for binding of $O_2$. Although $O_2$ binding is a complex process that involves structural changes in the protein, it can be approximated per monomer as the simple equilibrium process:

$$Hm + O_2(g) \rightleftharpoons HmO_2$$

## 4. Add a Catalyst

The addition of a catalyst increases the rate of a reaction by decreasing the activation energy. But there is also a decrease for the activation energy of the reaction in the opposite direction. The net result is that there is no impact on the position of the equilibrium (or the value of $K$). The reaction simply establishes equilibrium faster.

## THIS LAB

The goal in this lab is to study the impact of reactant/product concentrations and temperature changes on established chemical and physical equilibria. This will be accomplished by setting up a series of chemical equilibria and then adding various reagents, or changing the temperature, followed by characterizing the changes in the equilibria. The equilibria will be monitored in two different ways.

# 1. Colorimetric Changes

It is possible to estimate the position of some equilibria (those that contain colored species) by noting the color of the solution. If more than one species is present in the same solution, then the color of the solution will be a combination of the colors of the various species present. The first two equilibria in this lab contain **metal complexes.** A metal complex contains a central metal that interacts with multiple ions/molecules (called **ligands**) through coordinate covalent bonds (i.e., a bond where the ligand donates both electrons to form a covalent interaction with the metal). Metal complexes can be converted to other complexes by exchanging the type of ligands interacting with the metal—this often results in a color change as observed in this lab. For each reaction, an initial equilibrium will be established. Then, reactant and/or product concentrations will be changed, or the temperature will be adjusted, to introduce a disruption to the equilibrium. There will be three equilibria studied in this section:

*Iron(III) and thiocyanate:*

$$Fe(H_2O)_6^{3+}(aq) + SCN^-(aq) \rightleftharpoons Fe(H_2O)_5(SCN)^{2+}(aq) + H_2O(l)$$

gold                                     dark red-brown

*Cobalt(II) and chloride:*

$$Co(H_2O)_6^{2+} + 4Cl^-(aq) \rightleftharpoons CoCl_4^{2-} + 6H_2O(l)$$

pink                              blue
octahedral geometry          tetrahedral geometry

*Bromophenol blue (HBb):*

$$HBb(aq) \rightleftharpoons H^+(aq) + Bb^-(aq)$$

yellow                         purple

$CoCl_2$ is a colorimetric indicator when added to the commercial drying agent Drierite. Drierite contains ~98% $CaSO_4$ and ~2% $CoCl_2$. The $CaSO_4$ absorbs moisture from the surroundings, while the $CoCl_2$ causes the Drierite to appear blue when no moisture has been absorbed and turn pink in the presence of water vapor. In the third equilibrium, bromophenol blue, a weak acid used as an acid-base indicator, will be the focus of study. Bromophenol blue changes from yellow at pH 3.0 to purple at pH 4.6.

# 2. Solubility Changes in Precipitation Reactions

Reactions that involve the formation/dissolution of a solid are also equilibrium processes, where the solid precipitate is in equilibrium with a saturated solution of the constituent ions. In this lab, you will be studying the sparingly soluble compound calcium hydroxide:

$$Ca(OH)_2(s) \rightleftharpoons Ca^{2+}(aq) + 2OH^-(aq)$$

After initially establishing the equilibrium, stresses will be introduced to understand Le Châtelier's principle.

## *Procedure*

For Parts I and II, you will prepare a series of reaction mixtures. Take a test tube rack and place ~15 small test tubes in the rack—these tubes should be clean and dry. Label the test tubes "1" through "15". You will also need to prepare both a hot water bath and an ice water bath. For the hot water bath, fill a 250-mL beaker approximately two-thirds full with water (can be from the tap) and begin heating on a hot plate—*do not let the water boil away.* For the ice water bath, fill a 250-mL beaker to the top with crushed ice. Then add enough water to make an ice water slush.

## I.  Colorimetric Changes to Monitor Equilibrium Shifts

### Iron(III) and Thiocyanate

$$Fe(H_2O)_6^{3+}(aq) + SCN^-(aq) \rightleftharpoons Fe(H_2O)_5(SCN)^{2+}(aq) + H_2O(l)$$

gold                                    dark red-brown

#### A.  Set Up the Equilibrium

1.  Use deionized water to fill test tube 1 two-thirds full.
2.  To this tube, add 12 drops of a 0.1 $M$ Fe(NO$_3$)$_3$ stock solution and 5 drops of a 0.1 $M$ KSCN solution. Mix thoroughly by holding the test tube securely with one hand and flicking or tapping the bottom of the tube with the fingers of your other hand.
3.  The solution color should be red-orange; if not, then repeat step 2 until this color is obtained.
4.  Distribute this solution evenly into four additional tubes, giving you five total tubes with the same volume.

#### B.  Make Changes to Equilibrium

5.  Test tube 1 will be used as a control—do nothing to the contents of this tube.
6.  Add 2 drops of 0.1 $M$ Fe(NO$_3$)$_3$ stock solution to tube 2 and mix well. Record what happens—when recording these observations pay close attention to color changes (as compared to tube 1), formation or dissolution of a precipitate, production of a gas, heating or cooling of the solution, and so on. Be sure your recorded observations are clear and precise.
7.  Add 2 drops of 0.1 $M$ KSCN solution to tube 3 and mix well; record observations.
8.  Add 3 to 5 drops of 0.1 $M$ AgNO$_3$ solution to tube 4 and mix well; record observations.
9.  Using a test tube clamp, place tube 5 into a hot water bath. After the tube contents have heated up (~3 to 5 min), remove the tube from the bath and record your observations.

### Cobalt(II) and Cl$^-$

$$Co(H_2O)_6^{2+} + 4Cl^-(aq) \rightleftharpoons CoCl_4^{2-} + 6H_2O(l)$$

pink                                    blue

octahedral geometry          tetrahedral geometry

#### A.  Set Up the Equilibrium

10.  Fill tube 6 two-thirds full with the CoCl$_2$ in ethanol solution provided by your instructor. Record your observations on this initial solution.
11.  Add just enough deionized water dropwise until a color change occurs. Record your observations.
12.  Distribute this solution evenly into two additional tubes (7 and 8), giving you three total tubes with the same volume.

#### B.  Make Changes to Equilibrium

13.  Test tube 6 will be used as a control—do nothing to the contents of this tube.
14.  Add a concentrated HCl solution dropwise to tube 7 until a color change occurs. Record your observations.

15. Using a test tube clamp, place tube 8 into a hot water bath. After the tube contents have heated up (~2 to 5 min), remove the tube from the bath and record your observations.
16. Now place tube 8 into an ice water bath. After the tube contents have cooled (~2 to 5 min), remove the tube from the bath and record your observations.

## Bromophenol Blue (HBb)

$$HBb(aq) \rightleftharpoons H^+(aq) + Bb^-(aq)$$

yellow                                  purple

### A. Set Up the Equilibrium

17. Use deionized water to half fill tubes 9, 10, and 11.
18. To each of these tubes, add three drops of bromophenol blue stock solution; mix well.

### B. Make Changes to Equilibrium

19. Test tube 9 will be used as a control—do nothing to the contents of this tube.
20. Add 2 drops of 0.1 $M$ HCl to tube 10; mix well. Record observations.
21. Add 2 drops of 0.1 $M$ NaOH to tube 11; mix well. Record observations.
22. Slowly add 5 drops of 0.1 $M$ NaOH to tube 10; mix well. Record observations of what happened as you add base.
23. Dispose of all samples as instructed by your instructor—do not pour material in the sink. Rinse with deionized water, then shake each tube out to remove as much water as possible. Place tubes back into your tube rack.

## II.  Solubility Changes That Reflect Equilibrium Shifts

## Calcium hydroxide

$$Ca(OH)_2(s) \rightleftharpoons Ca^{2+}(aq) + 2OH^-(aq)$$

### A. Set Up the Equilibrium

24. Measure ~0.2 g of Ca(OH)$_2$ and place in a 250-mL Erlenmeyer flask. Add ~100 mL of deionized water to the flask and mix. Note that all of the solid will not dissolve.
25. Allow the solution to sit for 5 min, then isolate the clear filtrate above the solid (this is a saturated solution) by filtering the solution to collect the filtrate.
26. Place 5 mL of the filtrate into a medium sized test tube.

### B. Make Changes to Equilibrium

27. To the test tube, add 5 mL of a 6 $M$ NaOH solution and mix thoroughly. Record your observations.
28. Now add 5 mL of a 1.0 $M$ CaCl$_2$ solution and mix thoroughly. Record your observations.
29. Finally, add 8 mL of a 6 $M$ HCl solution and mix thoroughly. Record your observations.

### I. Colorimetric Changes to Monitor Equilibrium Shifts

#### Iron(III) and Thiocyanate

1. Color of the initial solutions:

   $Fe(NO_3)_3$ solution       _____

   KSCN solution       _____

2. Observations for tube 1:     _____

3. Observations for tube 2:     _____

4. Observations for tube 3:     _____

5. Observations for tube 4:     _____

6. Observations for tube 5     _____

#### Cobalt(II) and Cl⁻

7. Color of the initial solution:

   $CoCl_2$ solution (tube 6)       _____

8. Observations for tube 6 ($+H_2O$):     _____

9. Observations for tube 7:     _____

10. Observations for tube 8 (hot water):     _____

11. Observations for tube 8 (ice water):     _____

#### Bromophenol Blue (HBb)

12. Color of the initial solutions:

    HBb solution (tube 9):       _____

13. Observations for tube 10 ($+HCl$):     _____

14. Observations for tube 11:     _____

15. Observations for tube 10 ($+NaOH$):     _____

### II. Solubility Changes That Reflect Equilibrium Shifts

16. Observations for addition of NaOH:     _____

17. Observations for addition of $CaCl_2$:     _____

18. Observations for addition of HCl:     _____

## *Analysis*

### I. Colorimetric Changes to Monitor Equilibrium Shifts

#### Iron(III) and Thiocyanate

Based on your observations, which direction did the equilibrium shift (if at all) when the following were done:

**19.** Addition of $Fe(NO_3)_3$ to tube 2: _____

**20.** Addition of KSCN to tube 3: _____

**21.** Addition of $AgNO_3$ to tube 4: _____

**22.** Heating of tube 5: _____

#### Cobalt(II) and Cl⁻

Based on your observations, which direction did the equilibrium shift (if at all) when the following were done:

**23.** Addition of $H_2O$ to tube 6: _____

**24.** Addition of HCl to tube 7: _____

**25.** Heating of tube 8: _____

**26.** Cooling of tube 8: _____

#### Bromophenol Blue (HBb)

Based on your observations, which direction did the equilibrium shift (if at all) when the following were done:

**27.** Addition of HCl to tube 10: _____

**28.** Addition of NaOH to tube 11: _____

**29.** Addition of NaOH to tube 10: _____

### II. Solubility Changes That Reflect Equilibrium Shifts

Based on your observations, which direction did the equilibrium shift (if at all) when the following were done:

**30.** Addition of NaOH to tube: _____

**31.** Addition of $CaCl_2$ to tube: _____

**32.** Addition of HCl to tube: _____

## Reflection Questions

Questions 1 and 2 refer to the chemical equilibrium:

$$Fe(H_2O)_6^{3+}(aq) + SCN^-(aq) \rightleftharpoons Fe(H_2O)_5(SCN)^{2+}(aq) + H_2O(l)$$

1. Explain in terms of Le Châtelier's principle your answer in
   (a) #19 (Analysis section):


   (b) #20 (Analysis section):


   (c) #21 (Analysis section):


2. (a) Explain in terms of Le Châtelier's principle your answer in #22 (Analysis section):


   (b) Based on (a), is the equilibrium an endothermic or exothermic process (or are you unable to tell)?


   (c) Does the value of $K$ change in (a)? If so, does it increase or decrease?


Questions 3 and 4 refer to the chemical equilibrium:

$$Co(H_2O)_6^{2+} + 4Cl^-(aq) \rightleftharpoons CoCl_4^{2-} + 6H_2O(l)$$

3. Explain in terms of Le Châtelier's principle your answer in
   (a) #23 (Analysis section):

(b) #24 (Analysis section):

4. (a) Explain in terms of Le Châtelier's principle your answers in #25 and #26 (Analysis section):

(b) Based on (a), is the equilibrium an endothermic or exothermic process (or are you unable to tell)?

(c) Does the value of $K$ change in (a)? If so, does it increase or decrease?

5. For the following reaction:

$$Ca(OH)_2(s) \rightleftharpoons Ca^{2+}(aq) + 2OH^-(aq)$$

Explain in terms of Le Châtelier's principle your answer in
(a) #30 (Analysis section):

(b) #31 (Analysis section):

(c) #32 (Analysis section):

6. The protein hemoglobin (abbreviated Hm) is the active species in your red blood cells that binds molecular oxygen and forms the complex $HmO_2$ (see Figure 1). Carbon monoxide is a poisonous agent that also binds to hemoglobin, displacing oxygen (i.e., CO binds preferentially to $O_2$). This can be represented as an equilibrium:

$$HmO_2 + CO \rightleftharpoons HmCO + O_2$$

To treat CO poisoning, (1) the person is removed from the high CO area and (2) $O_2$ is administered. Based on your understanding of equilibrium, explain how each of these steps helps treat the problem:

## Connection

Based on your experience in this lab, draw a connection to something in your everyday life or the world around you (something not mentioned in the background section):

# 20

# Titration II. pH Titration Curves

## Objectives

- Understand the techniques and equipment associated with titrations.
- Interpret titration curves for both weak and strong acids titrated with a strong base.

## Materials Needed

- 50-mL burette
- Burette stand
- Funnel
- 0.1000 $M$ sodium hydroxide (NaOH) standard solution (actual concentration to be provided)
- 0.2 $M$ hydrochloric acid (HCl)
- 0.2 $M$ acetic acid ($CH_3COOH$)
- Deionized water
- pH meter or Vernier pH electrode
- pH 4 buffer solution
- pH 10 buffer solution
- Phenolphthalein
- 250-mL Erlenmeyer flask(s)
- 10-mL volumetric pipette

## Safety Precautions

- Safety goggles/glasses should be worn during this lab.
- NaOH is a strong base that can cause burns. Avoid contact with body tissues.
- HCl is a strong acid that can cause burns. Avoid contact with body tissues.

## Background

In Laboratory 10, you learned about the equipment and procedures associated with acid/base titrations, an important volumetric analytical technique. It is recommended that the background section of that lab be reviewed prior to this experiment.

## THIS LAB

Acids behave in characteristic fashion upon titration with a strong base, depending on whether the acid is classified as either strong or weak. In this lab, you will generate two titration curves involving a representative strong acid (hydrochloric acid) and a representative weak acid (acetic acid) using a standard sodium hydroxide solution as the titrant. A standard solution is one in which the concentration has been accurately determined. The actual concentration of the standard sodium hydroxide solution used in this experiment will be provided by your instructor. Careful attention to technique during the titrations is critical for the preparation of smooth pH titration curves, which, in turn, will be needed for data analysis.

## Procedure

### CALIBRATION OF THE pH METER

1. As per the instructions provided for the pH meter, calibrate using pH 4 and 10 buffer solutions. Between measurements, store the calibrated pH probe in the pH 4 buffer or other solution as directed by your instructor.

### TITRATION OF A 0.2 $M$ HCl SOLUTION

2. Record the actual concentration of the 0.1000 $M$ NaOH standard solution.
3. Using a 10-mL volumetric pipette, transfer 10.00 mL of 0.2 $M$ HCl to a 250-mL Erlenmeyer flask.
4. Add 50 mL of deionized water and two to three drops of the phenolphthalein indicator.
5. Using a funnel, rinse a clean 50-mL burette with three 5-mL portions of the 0.1000 $M$ NaOH solution, making sure to pass the solution through the burette tip. Dispose of the waste NaOH solution in the appropriately labeled waste container.
6. Fill the burette with the NaOH standard solution using a funnel. The solution level should allow for at least 25 mL of added base without having to refill the burette. Check that no air bubbles are trapped in the burette tip. Remove the funnel and record the initial volume of the solution to the correct precision that the burette allows.
7. Calculate and record the volume of base solution expected to reach the endpoint of your titration.
8. Titrate the HCl solution by adding the NaOH solution with swirling of the Erlenmeyer flask to ensure complete and timely mixing of the two solutions. The tip of the burette should be below the rim of the Erlenmeyer flask to prevent the loss of titrant. Initially, add the NaOH solution in approximately 1-mL amounts, pausing after each addition to allow for complete mixing of the solution and recording of the pH using the pH meter. After each addition, record the volume of base in the burette, the pH of the solution, and the color of the indicator using the appropriate data table.

   *Note:* When the volume of added base is between 1 and 2 mL less than calculated to reach the endpoint, add the NaOH solution in *much* smaller increments (near the endpoint, 0.1-mL additions are appropriate). Record the data, paying attention to the color of the indicator after each addition. As the endpoint of the titration nears, the pink color of the indicator will become more persistent, necessitating very small additions of NaOH. Good technique and monitoring of the titration experiment through observation will allow for more data to be collected near the endpoint of the titration, which will, in turn, provide a smoother titration curve for analysis. Continue with smaller additions until approximately 1 mL of base has been added beyond the endpoint of the titration, then add in 1-mL portions for the addition of a final 5 to 10 mL of base.

9. Within the lab period, convert the burette readings in your data table (the raw data) to the actual volume of base added and then plot the pH of the solution versus the volume of NaOH added. On your plot, clearly indicate the endpoint of the titration. If your titration curve does not reflect a smooth transition of points, in particular near the endpoint, then repeat this portion of the lab and modify your acquisition of data accordingly.

### TITRATION OF A 0.2 $M$ CH$_3$COOH SOLUTION

10. Using a 10-mL volumetric pipette, transfer 10.00 mL of 0.2 $M$ CH$_3$COOH to a 250-mL Erlenmeyer flask.
11. Add 50 mL of deionized water and two to three drops of the phenolphthalein indicator.

12. Titrate the $CH_3COOH$ solution using the NaOH standard solution as described in step 8.
13. Convert the burette readings in your data table to the volume of base added and then plot the pH of the solution versus the volume of NaOH added. On your plot, clearly indicate the endpoint of the titration, the region of maximum buffering, and the point on the titration curve where the $pH = pK_a$. If your titration curve does not reflect a smooth transition of points, in particular near the endpoint, then repeat this portion of the lab and modify your acquisition of data accordingly.

## Data Collection

### ACTUAL CONCENTRATION OF THE STANDARD 0.1 M NaOH SOLUTION

_____

### TITRATION OF A 0.2 M HCl SOLUTION

1. Phenolphthalein indicator color: _____ (acidic form) _____ (basic form)

2. Initial volume of NaOH solution, $V_i$ (initial burette reading, step 6): _____

3. Volume of base solution expected to reach the endpoint (step 7, show calculation and include a balanced chemical equation):

_____

4. Data table: $HCl(aq) + NaOH(aq)$

| Volume of base in burette (mL) | Volume of added base (mL) | pH | Indicator color |
|---|---|---|---|
| | 0.00 | | |
| | | | |
| | | | |
| | | | |
| | | | |
| | | | |
| | | | |
| | | | |
| | | | |
| | | | |
| | | | |

| Volume of base in burette (mL) | Volume of added base (mL) | pH | Indicator color |
|---|---|---|---|
|  |  |  |  |
|  |  |  |  |
|  |  |  |  |
|  |  |  |  |
|  |  |  |  |
|  |  |  |  |
|  |  |  |  |
|  |  |  |  |
|  |  |  |  |
|  |  |  |  |
|  |  |  |  |
|  |  |  |  |
|  |  |  |  |
|  |  |  |  |
|  |  |  |  |
|  |  |  |  |
|  |  |  |  |
|  |  |  |  |
|  |  |  |  |
|  |  |  |  |
|  |  |  |  |
|  |  |  |  |
|  |  |  |  |
|  |  |  |  |
|  |  |  |  |
|  |  |  |  |
|  |  |  |  |
|  |  |  |  |
|  |  |  |  |
|  |  |  |  |
|  |  |  |  |
|  |  |  |  |
|  |  |  |  |
|  |  |  |  |
|  |  |  |  |
|  |  |  |  |
|  |  |  |  |
|  |  |  |  |
|  |  |  |  |
|  |  |  |  |
|  |  |  |  |
|  |  |  |  |
|  |  |  |  |
|  |  |  |  |
|  |  |  |  |
|  |  |  |  |
|  |  |  |  |
|  |  |  |  |
|  |  |  |  |
|  |  |  |  |

**5.** Data table (**if necessary**): HCl($aq$) + NaOH($aq$)

| Volume of base in burette (mL) | Volume of added base (mL) | pH | Indicator color |
|---|---|---|---|
| | 0.00 | | |
| | | | |
| | | | |
| | | | |
| | | | |
| | | | |
| | | | |
| | | | |
| | | | |
| | | | |
| | | | |
| | | | |
| | | | |
| | | | |
| | | | |
| | | | |
| | | | |
| | | | |
| | | | |
| | | | |
| | | | |
| | | | |
| | | | |
| | | | |
| | | | |
| | | | |
| | | | |
| | | | |
| | | | |
| | | | |
| | | | |
| | | | |
| | | | |
| | | | |
| | | | |
| | | | |
| | | | |
| | | | |
| | | | |
| | | | |
| | | | |
| | | | |
| | | | |
| | | | |
| | | | |
| | | | |
| | | | |
| | | | |

| Volume of base in burette (mL) | Volume of added base (mL) | pH | Indicator color |
|---|---|---|---|
| | | | |
| | | | |
| | | | |
| | | | |
| | | | |
| | | | |
| | | | |
| | | | |
| | | | |
| | | | |
| | | | |
| | | | |
| | | | |

## TITRATION OF A 0.2 *M* CH₃COOH SOLUTION

**6.** Phenolphthalein indicator color: _____ (acidic form) _____ (basic form)

**7.** Initial volume of NaOH solution, $V_i$ (initial burette reading): _____

**8.** Volume of base solution expected to reach the endpoint (show calculation and include a balanced chemical equation):

_____

**9.** Data table: $CH_3COOH(aq) + NaOH(aq)$

| Volume of base in burette (mL) | Volume of added base (mL) | pH | Indicator color |
|---|---|---|---|
| | 0.00 | | |
| | | | |
| | | | |
| | | | |
| | | | |
| | | | |
| | | | |
| | | | |
| | | | |
| | | | |
| | | | |

| Volume of base in burette (mL) | Volume of added base (mL) | pH | Indicator color |
|---|---|---|---|
| | | | |
| | | | |
| | | | |
| | | | |
| | | | |
| | | | |
| | | | |
| | | | |
| | | | |
| | | | |
| | | | |
| | | | |
| | | | |
| | | | |
| | | | |
| | | | |
| | | | |
| | | | |
| | | | |
| | | | |
| | | | |
| | | | |
| | | | |
| | | | |
| | | | |
| | | | |
| | | | |
| | | | |
| | | | |
| | | | |
| | | | |
| | | | |
| | | | |
| | | | |
| | | | |
| | | | |
| | | | |
| | | | |
| | | | |
| | | | |
| | | | |
| | | | |
| | | | |
| | | | |
| | | | |
| | | | |
| | | | |

**10.** Data table (**if necessary**): $CH_3COOH(aq) + NaOH(aq)$

| Volume of base in burette (mL) | Volume of added base (mL) | pH | Indicator color |
|---|---|---|---|
| | 0.00 | | |
| | | | |
| | | | |
| | | | |
| | | | |
| | | | |
| | | | |
| | | | |
| | | | |
| | | | |
| | | | |
| | | | |
| | | | |
| | | | |
| | | | |
| | | | |
| | | | |
| | | | |
| | | | |
| | | | |
| | | | |
| | | | |
| | | | |
| | | | |
| | | | |
| | | | |
| | | | |
| | | | |
| | | | |
| | | | |
| | | | |
| | | | |
| | | | |
| | | | |
| | | | |
| | | | |
| | | | |
| | | | |
| | | | |
| | | | |
| | | | |
| | | | |
| | | | |
| | | | |
| | | | |
| | | | |
| | | | |

| Volume of base in burette (mL) | Volume of added base (mL) | pH | Indicator color |
|---|---|---|---|
| | | | |
| | | | |
| | | | |
| | | | |
| | | | |
| | | | |
| | | | |
| | | | |
| | | | |
| | | | |
| | | | |
| | | | |

## Analysis

**11.** Titration curves (plots of pH vs. volume of bases added):

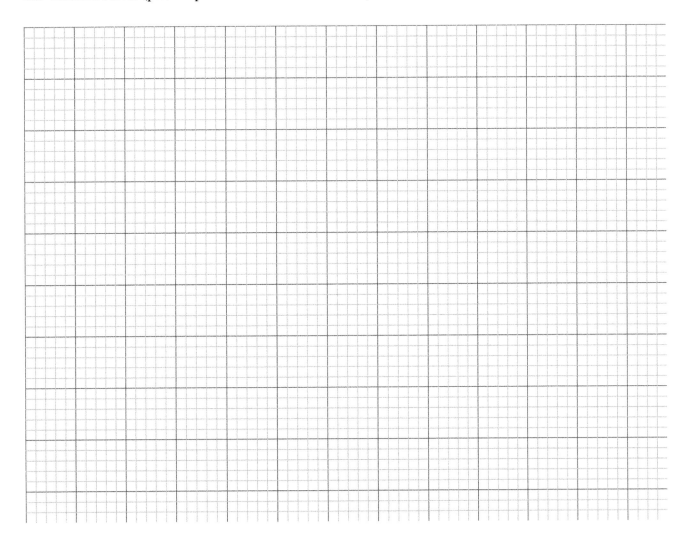

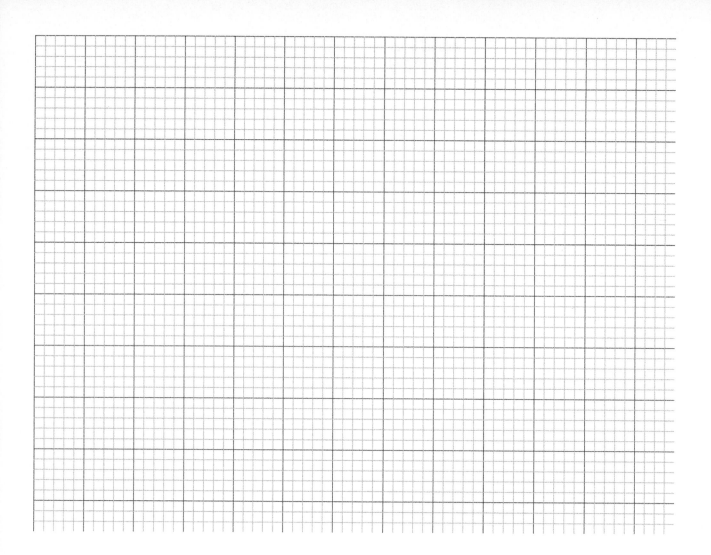

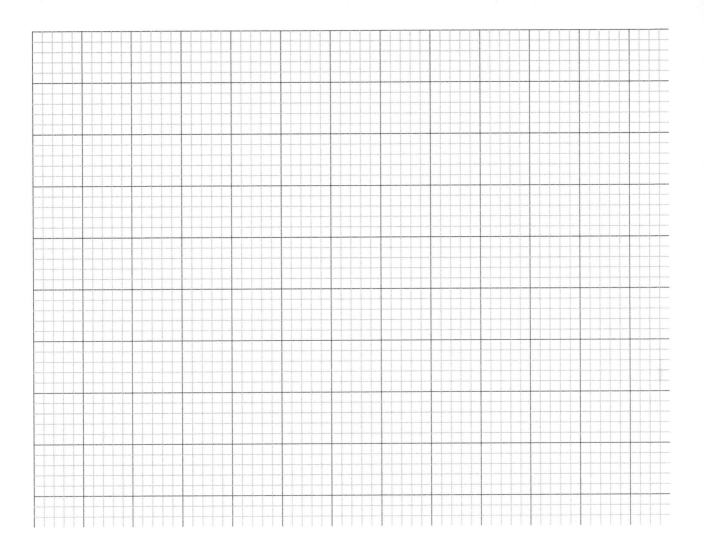

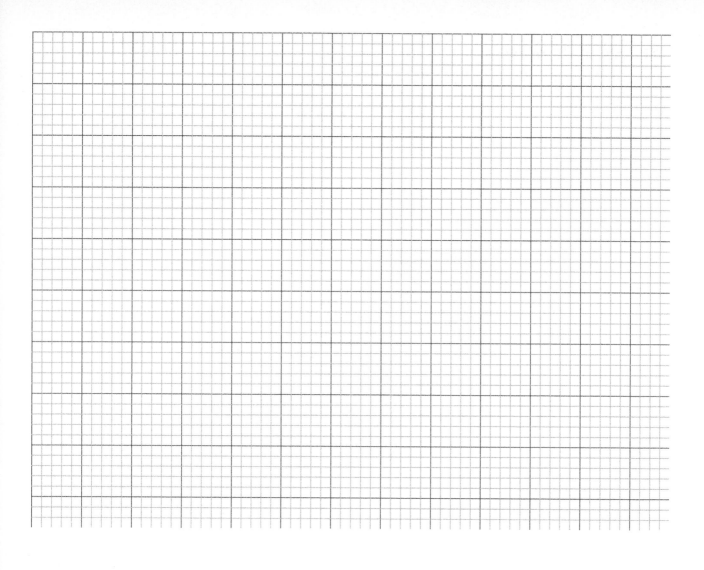

**12.** pH at the equivalence point for the titration of HCl(*aq*) (from titration curve):

_____

**13.** pH at the equivalence point for the titration of CH₃COOH(*aq*) (from titration curve):

_____

**14.** p$K_a$ of CH₃COOH (from titration curve):

_____

**15.** $K_a$ of CH₃COOH (experimental value):

_____

**16.** % error $= \dfrac{|(\text{known } K_a \text{ for weak acid}) - (\text{calculated } K_a \text{ for weak acid})|}{(\text{known } K_a \text{ for weak acid})} \times 100$

_____

1. Draw the Lewis structure of acetic acid and identify the most acidic proton.

2. Provide a clear explanation for the difference in the pH at the equivalence points in the two titrations.

3. Is there an error associated with comparing your experimentally determined $K_a$ value with the known value for acetic acid?

4. Was phenolphthalein a good choice for determining the endpoint of each titration? Suggest an indicator that would work uniquely for the strong acid/strong base titration in this lab (would NOT work for the weak acid/strong base titration).

5. (a) From your data, determine the actual concentration of the 0.2 $M$ HCl solution.

(b) From your data, determine the actual concentration of the 0.2 $M$ CH$_3$COOH solution.

6. Using your answer to question 5a, calculate the pH of the hydrochloric acid solution at the following points during the titration. (*Note:* Do not use your titration curve to answer this question.)

(a) Starting point

(b) After 8.00 mL of base has been added

(c) At the equivalence point

7. Using your answer to question 5b, calculate the pH of the acetic acid solution at the following points during the titration. (*Note:* Do not use your titration curve to answer this question.)

(a) Starting point

(b) After 8.00 mL of base has been added

(c) At the equivalence point

Based on your experience in this lab, draw a connection to something in your everyday life or the world around you (something not mentioned in the background section):

# 21

# Determining the Molar Solubility Product of Copper(II) Tartrate

## Objectives

- Measure the molar solubility of a salt with limited solubility.
- Measure the solubility product constant for a salt with limited solubility.
- Measure the molar solubility in the presence of a common ion.

## Materials Needed

- Deionized water
- 0.2 $M$ copper(II) sulfate ($CuSO_4$) solution (prepared by instructor)
- 0.2 $M$ sodium tartrate ($Na_2C_4H_4O_6$) solution (prepared by instructor)
- 0.1 $M$ sodium sulfate ($Na_2SO_4$) solution (prepared by instructor)
- 0.1 $M$ sulfuric acid ($H_2SO_4$) solution (prepared by instructor)
- 250-mL beaker (1)
- 50-mL Erlenmeyer flask (1)
- Hot plate
- Filter paper and funnel (or centrifuge and centrifuge tubes)
- Medium test tubes (3)
- Test tubes or scintillation vials (5)
- Spatula
- Parafilm
- Glassware to collect filtrates (supernatants if using a centrifuge)
- Method for accurately measuring the absorbance of samples at a wavelength of 635 nm. Possible options include:
  - Vernier LabPro handheld device with a SpectroVis Plus spectrophotometer attachment. The associated software LoggerPro must be installed on a computer to allow downloading and manipulation of data. Information about the Vernier device and associated peripherals and software can be obtained from the website http://www.vernier.com (last accessed May 2011).
  - UV/Vis spectrophotometer
  - Spectronic 20 spectrophotometer
- Cuvettes compatible with the device used to measure absorbance

## Safety Precautions

- Gloves and safety goggles/glasses should be worn during this lab.
- Sulfuric acid is highly corrosive. Avoid contact with skin or eyes, or ingestion.
- Dispose of waste solutions in designated waste containers as indicated by your instructor.

## Background

As demonstrated by the "solubility rules," the solubility of ionic solids varies greatly. For example, the Group IA salts, such as NaCl and KBr, are very soluble, while many phosphate, carbonate, and hydroxide salts are poorly soluble. However, the categorization of ionic compounds as either "soluble" or "insoluble," as is commonly used in a first-semester general chemistry course, is a simplistic view of solubility. In fact, the *most soluble* ionic compounds become *insoluble* at the saturation limit of water; and the *most insoluble* ionic compounds are *soluble* if low concentrations of solute are present. A more thorough approach to understanding solubility requires a quantitative definition. Considering solubility as a gradient, where the "insoluble" ionic compounds lie at one extreme and the "soluble" ionic compounds lie at the other extreme, provides a convenient means to understand the relationship between equilibrium and solubility.

For ionic solids with limited solubility, the dissolution of the solid into water is treated by the following equilibrium:

$$A_xB_y(s) \rightleftarrows x\ A^{y+}(aq) + y\ B^{x-}(aq) \qquad \textbf{Reaction 1}$$

Upon placing solid $A_xB_y$ into water, the rate for the forward dissolution reaction is relatively large. As $[A^{y+}]$ and $[B^{x-}]$ increase over time, the rate of the reverse precipitation reaction increases. Equilibrium is established once the forward and reverse reaction rates are equivalent. This equilibrium is heterogeneous, with the solid in contact with a saturated solution of the constituent ions. Very soluble ionic compounds have equilibria that lie far to the products' side, while more insoluble compounds have equilibria biased to the reactant side of Reaction 1.

As with any equilibrium, an equilibrium constant expression can be written for Reaction 1:

$$K_c = [A^{y+}]^x[B^{x-}]^y = K_{sp} \qquad \textbf{Equation 1}$$

When the equilibrium involves the solubilization of a sparingly soluble ionic solid, $K_c$ is renamed $K_{sp}$ and is called the **solubility product constant.** As examples, the solubility equilibria and $K_{sp}$ expressions are shown for AgCl and $Ca_3(PO_4)_2$:

$$AgCl(s) \rightleftarrows Ag^+(aq) + Cl^-(aq) \qquad K_{sp} = [Ag^+][Cl^-]$$

$$Ca_3(PO_4)_2(s) \rightleftarrows 3Ca^{2+}(aq) + 2PO_4^{3-}(aq) \qquad K_{sp} = [Ca^{2+}]^3[PO_4^{3-}]^2$$

Two important points regarding solubility equilibria are (1) as is the case for any equilibrium expression, solids are not included in the $K_{sp}$ expression; and (2) by convention, the equilibrium reaction is always written with the ionic solid as the reactant and the solubilized ions as products.

When the value of $K_{sp}$ is known, one can determine the **molar solubility** of the ionic solid, where molar solubility is defined as the moles of solid that will dissolve to form one liter of saturated solution. For the AgCl and $Ca_3(PO_4)_2$ examples above, the molar solubility of each can be calculated as shown next using "ICE" tables:

Example 1. Calculate the molar solubility of AgCl:

$$AgCl(s) \rightleftharpoons Ag^+(aq) + Cl^-(aq)$$

| | | |
|---|---|---|
| Initial | $0\,M$ | $0\,M$ |
| Change | $+x$ | $+x$ |
| Equilibrium | $x$ | $x$ |

$$K_{sp} = 1.8 \times 10^{-10} = [Ag^+][Cl^-] = x^2 \qquad \textbf{Equation 2}$$

$$x = 1.3 \times 10^{-5}\,M = \text{molar solubility of AgCl}$$

Example 2. Calculate the molar solubility of $Ca_3(PO_4)_2$:

$$Ca_3(PO_4)_2(s) \rightleftharpoons 3Ca^{2+}(aq) + 2PO_4^{3-}(aq)$$

| | | |
|---|---|---|
| Initial | $0\,M$ | $0\,M$ |
| Change | $+3x$ | $+2x$ |
| Equilibrium | $3x$ | $2x$ |

$$K_{sp} = 1 \times 10^{-25} = [Ca^{2+}]^3[PO_4^{3-}]^2 = (3x)^3(2x)^2 = (27x^3)(4x^2) = 108x^5 \qquad \textbf{Equation 3}$$

$$x = 4 \times 10^{-6}\,M = \text{molar solubility of } Ca_3(PO_4)_2$$

It is worth noting that nothing is entered under the $AgCl(s)$ and $Ca_3(PO_4)_2(s)$ in their respective ICE tables since they do not appear in the $K_{sp}$ expressions (their amounts, but *not* their concentrations, change over time), and so the exact amount of solid is not important. However, in order for an equilibrium to be established, there must be enough solid present to form a saturated solution. If the $K_{sp}$ value is not known, but the molar solubility has been measured for a given ionic solid, then the $K_{sp}$ value can be easily calculated with the appropriate $K_{sp}$ equation (similar to Equation 2 or 3) by assigning the molar solubility value to $x$. For example, if it has been determined that the molar solubility of $Ag_2CrO_4(s)$ is $6.7 \times 10^{-5}\,M$ at 25 °C, the $K_{sp}$ value can be calculated as shown:

| $Ag_2CrO_4$ | $Ag_2CrO_4(s) \rightleftharpoons 2Ag^+(aq)$ | $+$ | $CrO_4^{2-}(aq)$ |
|---|---|---|---|
| Initial | $0\,M$ | | $0\,M$ |
| Change | $+2(6.7 \times 10^{-5}\,M)$ | | $+6.7 \times 10^{-5}\,M$ |
| Equilibrium | $2(6.7 \times 10^{-5}\,M)$ | | $6.7 \times 10^{-5}\,M$ |

$$K_{sp} = [Ag^+]^2[CrO_4^{2-}] = [2(6.7 \times 10^{-5}\,M)]^2(6.7 \times 10^{-5}\,M) = 1.2 \times 10^{-12} \qquad \textbf{Equation 4}$$

Molar solubilities for different ionic solids can be compared to predict relative solubilities. When the salts contain the same total number of ions, then the $K_{sp}$ values can be simply compared to predict relative solubilities (this is true since the form of the $K_{sp}$ equation is the same in each case). For example, when comparing the sparingly soluble ionic solids AgCl, $BaCrO_4$, and LiF, one can use their $K_{sp}$ values ($1.8 \times 10^{-10}$, $2.3 \times 10^{-10}$, and $5 \times 10^{-3}$, respectively) to determine that their relative solubilities in water are LiF > $BaCrO_4$ > AgCl. However, for the solids AgCl, $Ca_3(PO_4)_2$, and $Ag_2CrO_4$, one must use their molar solubilities ($1.3 \times 10^{-5}\,M$, $4 \times 10^{-6}\,M$, and $6.7 \times 10^{-5}\,M$, respectively) to determine that their relative solubilities in water are $Ag_2CrO$ > AgCl > $Ca_3(PO_4)_2$. Importantly, molar solubilities provide a quantitative means for reporting the solubility of any ionic compound, a more thorough treatment of solubility than the qualitative approach used by the solubility rules. In fact, all of the insoluble ionic compounds as determined from the solubility rules are actually better categorized as poorly soluble with finite solubilities in water governed by respective $K_{sp}$ values that can be determined using the preceding approaches.

## PREDICTING IF A PRECIPITATE WILL FORM

When dissolving an ionic compound in water, or when combining two solutions, each containing an ionic compound, one of three possible results will occur:

Result 1. An unsaturated solution will be formed; more solid could be added and dissolved.
Result 2. A saturated solution is formed; no more solid can be dissolved, any additional added solid remains undissolved.
Result 3. A supersaturated solution is formed in which more solute is in solution than what will be present at equilibrium; supersaturated solutions are typically formed at higher temperatures and are unstable; excess solute will precipitate from supersaturated solutions until a saturated solution is generated.

These results can be explained by remembering that solubility is governed by an equilibrium between a sparingly soluble salt and its dissolved ions:

$$A_xB_y(s) \rightleftarrows x\,A^{y+}(aq) + y\,B^{x-}(aq)$$

As with any equilibrium, a reaction quotient $Q_{sp}$ can be defined for the dissociation reaction:

$$Q_{sp} = [A^{y+}]^x[B^{x-}]^y$$

As described in Lab 18, the values of $Q$ and $K$ (in this case, $Q_{sp}$ and $K_{sp}$) can be compared to determine in which direction the reaction will proceed; that is, forward toward products (dissolution of the solid), or in reverse toward reactants (formation of a precipitate). So the three results can be defined as

Result 1. An unsaturated solution will be formed; more solid could be added and dissolved. **This will be true when $Q_{sp} < K_{sp}$.**
Result 2. A saturated solution is formed; no more solid can be dissolved, but no solid remains undissolved (and none precipitates from solution). **This will be true when $Q_{sp} = K_{sp}$, and the system is at equilibrium.**
Result 3. A supersaturated solution is formed (this is possible if two solutions are mixed); the result is that the solid will precipitate from solution until a saturated solution is generated, with the solid precipitate then in equilibrium with the solution. **This will be true when $Q_{sp} > K_{sp}$.**

So, given the concentrations of the dissolved ions, one can calculate $Q_{sp}$ and determine which direction the reaction will shift, and whether a precipitate will form (or if a given initial quantity of solid will dissolve). For example, if 50.0 mL of a 0.0010 $M$ $CaCl_2$ solution are mixed with 50.0 mL of a 0.010 $M$ $Na_2SO_4$ solution, will a $CaSO_4$ precipitate occur? To solve, one would determine the values for $[Ca^{2+}]$ and $[SO_4^{2-}]$, substitute those into the $Q_{sp}$ expression for $CaSO_4$ and then compare $Q_{sp}$ to $K_{sp}$:

$$[Ca^{2+}] = \frac{(0.0500\ L)(0.0010\ M)}{0.1000\ L} = 5.0 \times 10^{-4}\ M$$

$$[SO_4^{2-}] = \frac{(0.0500\ L)(0.010\ M)}{0.1000\ L} = 5.0 \times 10^{-3}\ M$$

$$Q_{sp} = (5.0 \times 10^{-4})(5.0 \times 10^{-3}) = 2.5 \times 10^{-6} \qquad K_{sp} = 2.4 \times 10^{-5}$$

$$Q_{sp} < K_{sp} \qquad \rightarrow \qquad \text{no precipitate forms}$$

# COMMON ION EFFECT AND SOLUBILITY

The addition of an ion that is already present in solution from the dissolution of an ionic solid will decrease the solubility of said ionic compound—this is called the **common ion effect.** Similarly, attempting to dissolve a sparingly soluble ionic solid in water that contains a common ion will decrease the amount of that solid that will dissolve. The presence of the common ion will shift the equilibrium toward the solid (reactant) side, suppressing the dissolution of the solid.

## EFFECT OF pH ON SOLUBILITY

The pH of a solution will have an impact on the solubility of an ionic solid if the constituent anion reacts with $H^+(aq)$ to form a weak acid. One obvious example involves hydroxide salts, such as $Mg(OH)_2$:

$$Mg(OH)_2(s) \rightleftharpoons Mg^{2+}(aq) + 2OH^-(aq)$$

For example, a decrease in pH [i.e., addition of $H^+(aq)$] will shift the equilibrium to the right, increasing the solubility of the solid [$H^+(aq)$ will react with $OH^-(aq)$, removing hydroxide from the equilibrium]. However, note that an increase in pH (i.e., addition of $OH^-$) will shift the equilibrium to the left, causing a formation of $Mg(OH)_2$ precipitate due to the common ion effect.

Another example of a salt whose solubility is impacted by pH is $BaF_2$:

$$BaF_2(s) \rightleftharpoons Ba^{2+}(aq) + 2F^-(aq)$$

$F^-$ is the conjugate base of HF, so a decrease in pH will remove $F^-$ from solution, shifting the equilibrium to the right and increasing the solubility of $BaF_2$.

## THIS LAB

The focus of this lab is to determine both the $K_{sp}$ and molar solubility of the ionic compound copper(II) tartrate $(CuC_4H_4O_6)$.[1] The tartrate ion (Figure 1) has a long history in terms of food and structural chemistry.

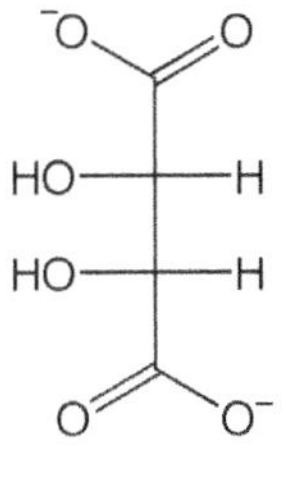

**Figure 1** **Structure of the tartrate ion.**

Potassium hydrogen tartrate $(KHC_4H_4O_6)$, also called cream of tartar, is used in baking as a source of acid—it reacts with bicarbonate to generate $CO_2(g)$:

$$HC_4H_4O_6^- + HCO_3^- \rightarrow C_4H_4O_6^{2-} + H_2CO_3$$

$$H_2CO_3 \rightarrow CO_2(g) + H_2O$$

$KHC_4H_4O_6$ is also found in grapes; known as "wine acid," it forms a precipitate during the fermentation of grapes. In fact, the primary source of $KHC_4H_4O_6$ for sale as cream of tartar is from the production of wine. The fully protonated form, tartaric acid $(H_2C_4H_4O_6)$, was studied by the French chemist and microbiologist Louis Pasteur. Tartaric acid, isolated from grapes, was found to rotate plane-polarized light to the right, while chemically synthesized tartaric acid (same chemical formula) did not rotate polarized light. Pasteur ultimately determined that tartaric acid comes in two forms that are identical in composition but exist as nonsuperimposable mirror images of each other. This initiated the study of molecular stereochemistry, a topic of considerable importance in organic and biochemistry with direct application in the pharmaceutical industry. Tartaric acid is used in candy, soft drinks, baked items, photography, and tanning.

In this lab, you will first prepare the $CuC_4H_4O_6$ salt by mixing copper(II) sulfate and sodium tartrate solutions:

$$CuSO_4(aq) + Na_2C_4H_4O_6(aq) \rightleftharpoons CuC_4H_4O_6(s) + Na_2SO_4(aq)$$

When the precipitate forms, the remaining solution will necessarily be saturated with $Cu^{2+}(aq)$ and $C_4H_4O_6^{2-}(aq)$:

$$CuC_4H_4O_6(s) \rightleftharpoons Cu^{2+}(aq) + C_4H_4O_6^{2-}(aq)$$

You will isolate the $CuC_4H_4O_6$ solid and redissolve it in water, sodium sulfate and sulfuric acid to determine the molar solubility under these conditions.

## References

1. Thomsen MW. "Determination of the Solubility Product of Copper(II) Tartrate." *Journal of Chemical Education.* 1992; 69(4): 328–329.
2. Dean JA. (ed.). *Lange's Handbook of Chemistry,* 13th Edition. McGraw-Hill: New York. 1985.

## Procedure

You will need to prepare a hot water bath. Fill a 250-mL beaker approximately two-thirds full with water (can be from the tap) and begin heating on a hot plate on the low to medium setting to get a temperature of 40°C to 50°C—DO NOT LET THE WATER BOIL AWAY.

### I.  Preparation of Solutions and Samples

*Note:* The following procedures involve filtration to isolate insoluble solids. An acceptable alternative would be to use a centrifuge, in which case the term "supernatant" replaces "filtrate" in this lab.

### Copper(II) Tartrate Solid

1. Place 10.0 mL of a 0.2 $M$ $CuSO_4$ solution and 10.0 mL of a 0.2 $M$ $Na_2C_4H_4O_6$ solution into a 50-mL Erlenmeyer flask and mix well. Set the solution aside for ~10 min to allow a precipitate to form, shaking the container periodically.
2. Filter the solution. Collect the $CuC_4H_4O_6$ solid for further studies.

### Copper(II) Tartrate Equilibrium Samples

3. Label three clean medium-size test tubes "1," "2," and "3."
4. Separate the $CuC_4H_4O_6$ solid from #2 into three approximately equal fractions, and place them into the three test tubes.
5. To tube 1, add ~6 mL of deionized water and mix well; all of the solid will not dissolve.
6. To tube 2, add ~6 mL of 0.1 $M$ $Na_2SO_4$ solution and mix well; all of the solid will not dissolve.
7. To tube 3, add ~6 mL of 0.1 $M$ $H_2SO_4$ solution and mix well; all of the solid will not dissolve.
8. Cover each of the tubes with a piece of Parafilm. Place the tubes into the hot water bath for ~10 min to aid with the establishment of equilibrium. Shake the tubes periodically.
9. Remove the tubes from the water bath, and allow them to cool for ~20 min.

10. Filter the contents of each of the tubes, collecting the filtrate—label the filtrate containers appropriately.

## Copper(II) Standards

11. Prepare a series of $Cu^{2+}$ standards by mixing the 0.2 $M$ $CuSO_4$ solution and deionized water in the proportions specified here using labeled, clean test tubes or scintillation vials:

| Standard solution # | Volume CuSO$_4$ (mL) | Volume deionized water (mL) |
|:---:|:---:|:---:|
| 1 | 3.00 | 1.00 |
| 2 | 2.00 | 3.00 |
| 3 | 2.00 | 8.00 |
| 4 | 2.00 | 13.00 |

## II. Generation of Copper(II) Calibration Curve

12. "Zero" the spectrophotometer using a deionized water blank.
13. Measure the absorbance at 635 nm for standard solutions 1 to 4 (#11). [*Note:* For each measurement, be sure the cuvette is clean, the outside of the cuvette is wiped clean with a piece of lens paper, and avoid bubbles (which will interfere with the measurement).]

## III. Measure Absorbance of Saturated Copper(II) Tartrate Solutions

14. Measure the absorbance at 635 nm for the following solutions:

    (1) filtrate from tube 1 (#5)

    (2) filtrate from tube 2 (#6)

    (3) filtrate from tube 3 (#7)

[*Note:* For each measurement, be sure the cuvette is clean, the outside of the cuvette is wiped clean with a piece of lens paper, and avoid bubbles (which will interfere with the measurement).]

## Data Collection

1. Absorbance at 635 nm for standard solutions:

    standard solution #1: _____

    standard solution #2: _____

    standard solution #3: _____

    standard solution #4: _____

2. Absorbance at 635 nm for $CuC_4H_4O_6$ filtrates:

    filtrate from tube 1 (#5): _____

    filtrate from tube 2 (#6): _____

    filtrate from tube 3 (#7): _____

## Analysis

3. Calculate the $[Cu^{2+}]$ in the standard solutions:

   (a) standard #1: _____

   (b) standard #2: _____

   (c) standard #3: _____

   (d) standard #4: _____

4. Plot $A_{635}$ vs. $[Cu^{2+}]$ for the four standard solutions—add a data point at $A_{635} = 0.0$, $[Cu^{2+}] = 0\ M$:

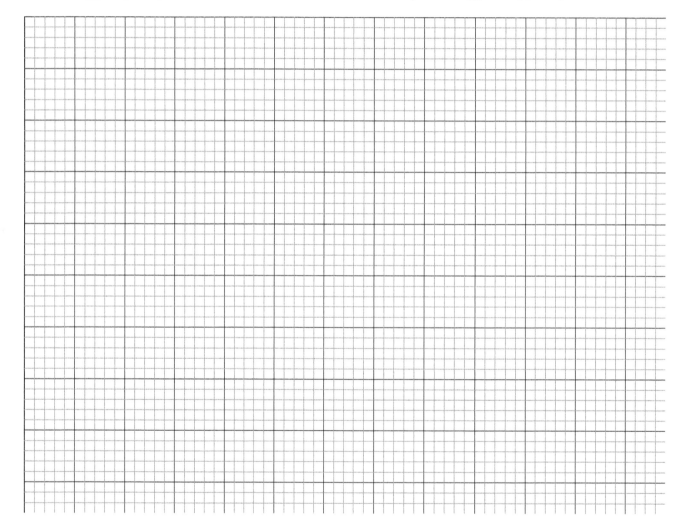

5. Using Excel, a graphing calculator, or the least-squares method, fit your data above to a straight line of the form:

$$A_{635} = m \times [Cu^{2+}] + (y\text{-intercept})$$

(a) Slope of your line (including units): _____

(b) $Y$-intercept of your line (including units): _____

6. Use the calibration line from #4 and #5 to determine the $[Cu^{2+}]$ in the following $CuC_4H_4O_6$ samples:

(a) filtrate from tube 1 (#5): _____

(b) filtrate from tube 2 (#6): _____

(c) filtrate from tube 3 (#7): _____

7. Calculate the $K_{sp}$ for $CuC_4H_4O_6$ from the filtrate from tube 1 (#5):

filtrate from tube 1 (#5): _____

8. Calculate the molar solubility of $CuC_4H_4O_6$ for the following samples

(a) filtrate from tube 1 (#5): _____

(b) filtrate from tube 2 (#6): _____

(c) filtrate from tube 3 (#7): _____

## Reflection Questions

1. In calculating $[Cu^{2+}]$ in the standard solutions (#3 in the Analysis section), what assumption(s) do you have to make?

   What data would be useful in confirming that your assumption is correct?

2. In plotting your standard line (#4 in the Analysis section), you were told to include a data point at $A_{635} = 0.0$, $[Cu^{2+}] = 0\ M$.
   (a) Explain why this is either a reasonable or unreasonable thing to do.

   (b) Does this extra data point significantly affect your calibration line? Demonstrate by using only the four measured data points from #3 (Analysis section) and generate a new calibration line, and compare this new calibration line to the one generated in the Analysis section (you could plot this new line on the sample graph as in #4).

   (c) Does the new calibration line (as compared to the one generated in the Analysis section) support your answer in part (a)? Explain.

3. In your original calibration line, what does the slope of the line represent?

4. In step 8 of the Procedure section, you are asked to cover each test tube with a piece of Parafilm and heat for ~10 min with periodic shaking.
   (a) What is the purpose of heating/shaking the samples?

   (b) Why is it important to cover the tubes with Parafilm?

   (c) What would be the possible impact on the values of $K_{sp}$ and molar solubility that you calculate from these samples if you did not heat? If you did not use Parafilm?

5. The literature value for the $K_{sp}$ of $CuC_4H_4O_6$ is $4 \times 10^{-4}$ (20 °C).[2] Calculate the percent error for your experimentally determined $K_{sp}$ values:

   (a) filtrate from tube 1 (#5): _____

   (b) What are the possible experimental errors that could explain your percent error?

6. Use Le Châtelier's principle to explain the molar solubilities that you calculated in #8b and c in the Analysis section (i.e., how do these values relate to those calculated in #8a and why?)

7. Suppose you had used twice as much $CuC_4H_4O_6(s)$ in #5, Procedure section. Would this increase, decrease, or have no effect on the value of $K_{sp}$ that you determined? Explain.

8. To generate the filtrates that you used to measure $A_{635}$, you had to filter the samples to remove the solid. If you did not filter the samples properly, what impact would that have on the measured $A_{635}$ and on the $[Cu^{2+}]$ that you would calculate?

## Connection

Based on your experience in this lab, draw a connection to something in your everyday life or the world around you (something not mentioned in the background section):

# 22

# Thermodynamics of Formation of a Borax Solution

## Objectives[1]

- Determine the value of the solubility product constant ($K_{sp}$) for the sparingly soluble salt borax at a variety of temperatures.
- Determine the values of $\Delta H°$, $\Delta G°$, and $\Delta S°$ for the reaction based on the $K_{sp}$ values.

## Materials Needed

- Deionized water
- Ice
- Sodium tetraborate decahydrate ($Na_2B_4O_7 \cdot 10H_2O$, or $Na_2[B_4O_5(OH)_4] \cdot 8H_2O$, also known as "borax")
- Bromocresol green indicator
- Standardized 0.500 $M$ HCl solution (prepared by instructor)
- 250-mL Erlenmeyer flask (1)
- 100-mL Erlenmeyer flask (4)
- 100-mL beaker (1)
- Large beaker
- 50-mL burette and burette stand
- Magnetic stir bar
- Hot plate
- 10-mL pipette
- Thermometer
- Ring stand with clamp to hold thermometer

## Safety Precautions

- Gloves and safety goggles/glasses should be worn during this lab.
- Dilute solutions of acid can cause irritations and burns. Avoid contact with skin and eyes, and avoid ingestion. Wash contaminated areas thoroughly with cold water.
- Hot glassware can cause burns; handle carefully.
- Dispose of waste solutions in designated waste containers as indicated by your instructor.

## Background

The spontaneity of a process depends on both enthalpic changes ($\Delta H$) as well as entropic changes ($\Delta S$). For chemical reactions, changes in enthalpy arise from the breakage and/or formation of chemical bonds and intermolecular forces, with a decrease in enthalpy (i.e., $\Delta H < 0$ or exothermic) contributing favorably to a spontaneous process. Entropy ($S$) is a measure of the disorder (randomness) of a system. A favorable change in entropy is one in which the disorder increases for the system ($\Delta S > 0$). Temperature is also important in determining the spontaneity of a process. When ice melts and forms liquid water, $\Delta H > 0$ (disruption of intermolecular forces between the water molecules) and $\Delta S > 0$ (solid-to-liquid conversion, representing an increase in disorder). At atmospheric pressure, this reaction is spontaneous above a temperature of 0°C but nonspontaneous at temperatures below 0°C.

These three factors ($\Delta H$, $\Delta S$, and temperature) can be combined in a new thermodynamic function, the Gibbs free energy ($G$, units = kJ mol$^{-1}$), based on the second law of thermodynamics:

$$\Delta G = \Delta H_{\text{system}} - T\,\Delta S_{\text{system}} \quad \text{(constant } T \text{ and } P)$$ **Equation 1**

The change in free energy can be used to predict the spontaneity of a process. At constant $T$ and $P$, a process is spontaneous *only* if $\Delta G < 0$. When $\Delta G > 0$, the process as written is not spontaneous, but the *reverse* process is spontaneous. Finally, when $\Delta G = 0$, the system is at equilibrium (i.e., neither direction is spontaneous). By looking at Equation 1 and the signs of $\Delta H$ and $\Delta S$, four distinct types of systems can be defined:

Type 1: $\Delta H < 0$ and $\Delta S > 0$: For this system, $\Delta G$ is *always* negative (i.e., at all temperatures $T$), and the system is *always* spontaneous. Examples of this type of system include

    a.  $2H_2O_2(aq) \rightarrow 2H_2O(l) + O_2(g)$

    b.  collapse of a building into a pile of bricks

Type 2: $\Delta H > 0$ and $\Delta S < 0$: For this system, $\Delta G$ is *always* positive (i.e., at all temperatures $T$), and the system is *never* spontaneous in the forward direction. Examples of this type of system include

    a.  $3O_2(s) \rightarrow 2O_3(g)$

    b.  a pile of bricks spontaneously assembling into a building

Type 3: $\Delta H > 0$ and $\Delta S > 0$: For this system, $\Delta G$ is negative (and the system is spontaneous) at temperatures *above* a specific value $T$. Examples of this type of system include

ice melting ; water evaporating

The specific value of $T$ in this case can be determined by setting $\Delta G = 0$ (equilibrium conditions):

$$\Delta G = 0 = \Delta H - T\,\Delta S$$

$$T = \frac{\Delta H}{\Delta S} \qquad\qquad (T \text{ in kelvins})$$

This is the temperature where equilibrium is established—above that temperature, the process is spontaneous, while below that temperature, the reverse process is favored. For example, at 1 atm, $T = 0$ K for the solid-to-liquid phase change, and $T = 100$ K for the liquid-to-gas phase change (phase changes are considered equilibria).

Type 4: $\Delta H < 0$ and $\Delta S < 0$: For this system, $\Delta G$ is negative (and the system is spontaneous) at temperatures *below* a specific value $T$ (value of $T$ determined as described above). Examples of this type of system include

water freezing ; water vapor condensing

## THIS LAB

In this lab, you will determine the thermodynamic quantities associated with the dissolution of the sparingly soluble solid sodium tetraborate decahydrate ($Na_2B_4O_7 \cdot 10H_2O$, or $Na_2[B_4O_5(OH)_4] \cdot 8H_2O$, also known as "borax") in water:

$$Na_2[B_4O_5(OH)_4] \cdot 8H_2O \rightleftharpoons 2Na^+(aq) + B_4O_5(OH)_4^{2-}(aq) + 8H_2O(l) \qquad \textbf{Reaction 1}$$

$$K_{sp} = [Na^+]^2[B_4O_5(OH)_4^{2-}] \qquad \textbf{Equation 2}$$

As is the case for all equilibria, the value of $K_{sp}$ (and thus the solubility of the sparingly soluble solid) is dependent on temperature (as seen in Lab 19). By measuring $K_{sp}$ for borax at various temperatures, the thermodynamic values $\Delta H°$, $\Delta G°$, and $\Delta S°$ can be determined.

Qualitatively, an increase in temperature for an endothermic reaction would make the process favor the products, and so $K_c$ would increase. In contrast, increasing the temperature for an exothermic reaction would lead to a decrease in $K_c$; the reaction would therefore becomes less product-favored at higher temperatures. This behavior can be quantified by beginning with Equations 3 and 4:

$$\Delta G_T° = -RT \ln K_T \qquad \textbf{Equation 3}$$
$$\Delta G_T° = \Delta H_T° - T\,\Delta S_T° \qquad \textbf{Equation 4}$$

where the subscript "$T$" indicates these equations can be used at different temperatures, and the superscript "°" indicates standard-state conditions (1 atm pressure, and typically 25°C, with the compounds in the reaction in their "standard states")—the standard states for pure substances are as follows:
gases = 1 atm pressure; liquids = pure liquid; solids = pure solid; elements = most stable allotrope under standard conditions; and solutions = 1 molar concentration.

One can then set Equations 3 and 4 equal to each other and rearrange:

$$-RT \ln K_T = \Delta H_T° - T\,\Delta S_T°$$

$$\ln K_T = \left(\frac{-\Delta H_T°}{R}\right)\left(\frac{1}{T}\right) + \left(\frac{\Delta S_T°}{R}\right) \qquad \textbf{Equation 5}$$

$\Delta H°$ and $\Delta S°$ are approximately constant with respect to temperature, so a plot of $\ln K_T$ vs. $\frac{1}{T}$ generates a straight line with a slope $= -\dfrac{\Delta H°}{R}$ and a y-intercept $= \dfrac{\Delta S°}{R}$. Equations 3 and 4 can then be used to determine $\Delta G°$.

In the borax equilibrium (Reaction 1), it can be seen that two sodium ions are generated in solution for each borate ion:

$$[Na^+] = 2[B_4O_5(OH)_4^{2-}] \qquad \textbf{Equation 6}$$

Substituting Equation 6 into Equation 2 yields an expression for $K_{sp}$ that depends only on $[B_4O_5(OH)_4^{2-}]$:

$$K_{sp} = 4[B_4O_5(OH)_4^{2-}]^3 \qquad \textbf{Equation 7}$$

Borate is a base, and so its concentration can be determined using a simple acid-base titration (as you performed in Lab 10). Each solution will be titrated with dilute standardized HCl solution to determine the borate ion concentration.

$$B_4O_5(OH)_4^{2-} + 2HCl + 3H_2O \rightarrow 4B(OH)_3 \quad [\text{or } H_3BO_3] + 2Cl^- \qquad \textbf{Reaction 2}$$

The endpoint will be signaled by the blue-to-yellow color change of the indicator bromocresol green.

This lab will be a group effort. For a lab of 30 students (15 pairs), five saturated borax solutions will be prepared at different temperatures (~5°C, ~20°C, ~35°C, ~45°C, and ~55 to 60°C). There will be multiple student pairs assigned to each temperature by the lab instructor. The results from each temperature will be averaged, giving all students 5 averaged values to perform their data analysis calculations with.

An interesting note: The largest borax mine in the world is located in Boron, California, a small community (~2250 people as of the 2010 census) named after the element boron and home of the U.S. Borax Boron Mine, an open-pit mine where borax is isolated as the ore colemanite. The mine supplies approximately half of the world's supply of refined borate compounds.[2] Borax has a number of uses: borax is used in household laundry and cleaning products; borate is used in buffers for biochemical applications such as gel electrophoresis; borax + $NH_4Cl$ is used for flux in welding iron and steel; in the Philippines, it is used for the extraction of gold in mining; it is used as a cross-linking agent to form the toy polymers Gak and Slime; it is a food additive in some Asian countries, but is banned in the United States; it is a fire retardant; it is an antifungal additive for fiberglass, and is used as an antifungal foot soak; it is an insecticide; it is used in fluoride detoxification; it is a curing agent for snake skins and salmon eggs; it is used as a buffering agent in swimming pools; boron in borax acts as a neutron absorber in nuclear reactors to control reactivity; and it generates a green tint in fire.

## References

1. The basic science in this lab—the use of measured $K$ values at different temperatures to determine thermodynamic quantities for an equilibrium—has appeared in numerous forms before [e.g., Mahan, Bruce H. "Temperature Dependence of Equilibrium: A First Experiment in General Chemistry." *J. Chem. Ed.* 1963; 40(6): 293–294, and Beaulieu, Lynn P. "A General Chemistry Thermodynamics Experiment." *J. Chem. Ed.* 1978; 55(1): 53–54].
2. "GPS for Machine Guidance Improves Safety at Borax Mine." *Mining Engineer* Dec. 2004; 21–23.

## Procedure

*Note: For this lab, work in pairs. For each temperature, there should be two or three groups collecting data, allowing an average borate concentration to be determined for that temperature. For best results, **DO NOT** transfer undissolved solid when removing a sample of the saturated solution for dilution and titration. Each student pair will titrate four solutions at a given temperature—be sure the four samples used for titrations come from the same original saturated borax solution.*

### I.  Preparation of Borax Equilibrium Solutions.

1. A large volume of deionized water should be placed in a large beaker and heated on a heating plate to ~45 to 50°C—two or three portions of this will be used later to rinse a graduated cylinder. Each group can prepare their own sample of warm deionized water, or several groups can share a large volume if limited heating plates are available. *Note:* (1) be sure this beaker does not go dry during the lab and (2) be sure to not contaminate this water if it is shared by multiple groups.
2. Measure ~30 g borax and place in a 250-mL Erlenmeyer flask containing a magnetic stir bar; add 75 mL deionized water.

   *For groups collecting data AT room temperature:*
3. Mix the solution with gentle stirring for at least 10 min. All of the solid should not dissolve—the presence of at least a small amount of undissolved solid is required to ensure that the solution is saturated.
4. Shut off the stirrer, insert a thermometer in the flask, and allow it to stand for at least 10 min more undisturbed to let any undissolved solid settle.
5. Proceed to step 16.

   *For groups collecting data BELOW room temperature:*
6. Place the Erlenmeyer flask in an ice-water bath and mix with gentle stirring for at least 20 min. All of the solid should not dissolve—the presence of at least a small amount of undissolved solid is required to ensure that the solution is saturated.

7. Shut off the stirrer, insert a thermometer in the flask, and allow it to stand for at least 10 min more undisturbed to let any undissolved solid settle.
8. Keep the flask in the ice-water bath until samples are removed for titration.
9. Proceed to step 16.

*For groups collecting data ABOVE room temperature:*
10. Place the Erlenmeyer flask on a heating plate; heat and stir the mixture gently until a temperature of ~10°C above your target temperature is reached (monitored with a thermometer suspended in the flask using a ring stand and an appropriate clamp).
11. If all of the borax solid dissolves, add a small additional amount of borax and continue heating and mixing. Continue adding small amounts of borax until a small amount of solid persists—the presence of at least a small amount of undissolved solid is required to ensure that the solution is saturated.
12. Maintain the temperature of the solution for at least 10 min, ensuring that some solid persists.
13. Turn off the heat and allow your mixture to cool to your target temperature (monitored with the thermometer). Additional precipitate will likely form as your solution cools—this is expected.
14. Try not to disturb the solid—your goal in the next step will be to collect *only* part of the saturated solution and *not* any of the solid.
15. Proceed to step 16.

*For ALL groups:*
16. Once the target temperature is reached, record the EXACT temperature.
17. CAREFULLY transfer 30.0 mL of the solution to a clean beaker, being sure to avoid transferring any solid to the beaker.
18. Use a pipette to transfer four 5.0-mL samples of this solution from the beaker into four separate Erlenmeyer flasks. Add to each flask 10.0 to 15.0 mL of deionizied water and 2 to 3 drops of bromocresol green indicator, which should turn the solutions blue. These will be the samples used for your titrations.
19. At this point, take a second to wash all glassware (EXCEPT THE 4 TITRATION SAMPLES), since borax can be difficult to remove from the glassware if it sits too long.

## II.  **Titration of Borax Solution Samples.**

20. Fill a clean 50-mL burette with the standardized 0.500 $M$ HCl solution.
21. Open the stopcock to allow a few milliliters of the solution to run through to remove air bubbles.
22. Using the titration methods you learned in Lab 10, titrate each of the four borate solutions to the appropriate endpoint (blue to yellow color transition). You might have to refill the burette with addition HCl. For each titration, be sure to record the initial and final volume readings on the burette.
23. When finished with the titrations, dispose of the excess HCl solution as indicated by your lab instructor and clean the burette.

## Data Collection

1. Temperature of the borax solutions at the point of removal from the flask: _____

2. Titration data:

|  | Trial 1 | Trial 2 | Trial 3 | Trial 4 |
|---|---|---|---|---|
| **Final burette reading (mL)** | | | | |
| **Initial burette reading (mL)** | | | | |
| **Volume HCl added to borax solution (mL)** | | | | |
| **Endpoint color** | | | | |

## *Analysis*

**3.** Use your titration data to determine the borate ion concentration in each trial:

Trial 1:                              [borate] = _____

Trial 2:                              [borate] = _____

Trial 3:                              [borate] = _____

Trial 4:                              [borate] = _____

**4.** Collect the titration and temperature data from the other groups:

| Temperature: ~5°C | | | | | | |
|---|---|---|---|---|---|---|
| **Group 1** | $temp_{actual}$ | | | **Group 2** | $temp_{actual}$ | |
| | $[borate]_{trial\ 1}$ | | | | $[borate]_{trial\ 1}$ | |
| | $[borate]_{trial\ 2}$ | | | | $[borate]_{trial\ 2}$ | |
| | $[borate]_{trial\ 3}$ | | | | $[borate]_{trial\ 3}$ | |
| | $[borate]_{trial\ 4}$ | | | | $[borate]_{trial\ 4}$ | |
| | | | | | | |
| **Group 3** | $temp_{actual}$ | | | | | |
| | $[borate]_{trial\ 1}$ | | | | | |
| | $[borate]_{trial\ 2}$ | | | | | |
| | $[borate]_{trial\ 3}$ | | | | | |
| | $[borate]_{trial\ 4}$ | | | | | |
| | | | | | | |
| **Temperature: ~20°C** | | | | | | |
| **Group 1** | $temp_{actual}$ | | | **Group 2** | $temp_{actual}$ | |
| | $[borate]_{trial\ 1}$ | | | | $[borate]_{trial\ 1}$ | |
| | $[borate]_{trial\ 2}$ | | | | $[borate]_{trial\ 2}$ | |
| | $[borate]_{trial\ 3}$ | | | | $[borate]_{trial\ 3}$ | |
| | $[borate]_{trial\ 4}$ | | | | $[borate]_{trial\ 4}$ | |
| | | | | | | |
| **Group 3** | $temp_{actual}$ | | | | | |
| | $[borate]_{trial\ 1}$ | | | | | |
| | $[borate]_{trial\ 2}$ | | | | | |
| | $[borate]_{trial\ 3}$ | | | | | |
| | $[borate]_{trial\ 4}$ | | | | | |
| | | | | | | |

| Temperature: ~35°C | | | | Group 2 | temp$_{actual}$ | |
|---|---|---|---|---|---|---|
| **Group 1** | temp$_{actual}$ | | | **Group 2** | temp$_{actual}$ | |
| | [borate]$_{trial 1}$ | | | | [borate]$_{trial 1}$ | |
| | [borate]$_{trial 2}$ | | | | [borate]$_{trial 2}$ | |
| | [borate]$_{trial 3}$ | | | | [borate]$_{trial 3}$ | |
| | [borate]$_{trial 4}$ | | | | [borate]$_{trial 4}$ | |
| | | | | | | |
| **Group 3** | temp$_{actual}$ | | | | | |
| | [borate]$_{trial 1}$ | | | | | |
| | [borate]$_{trial 2}$ | | | | | |
| | [borate]$_{trial 3}$ | | | | | |
| | [borate]$_{trial 4}$ | | | | | |

| Temperature: ~45°C | | | | | | |
|---|---|---|---|---|---|---|
| **Group 1** | temp$_{actual}$ | | | **Group 2** | temp$_{actual}$ | |
| | [borate]$_{trial 1}$ | | | | [borate]$_{trial 1}$ | |
| | [borate]$_{trial 2}$ | | | | [borate]$_{trial 2}$ | |
| | [borate]$_{trial 3}$ | | | | [borate]$_{trial 3}$ | |
| | [borate]$_{trial 4}$ | | | | [borate]$_{trial 4}$ | |
| | | | | | | |
| **Group 3** | temp$_{actual}$ | | | | | |
| | [borate]$_{trial 1}$ | | | | | |
| | [borate]$_{trial 2}$ | | | | | |
| | [borate]$_{trial 3}$ | | | | | |
| | [borate]$_{trial 4}$ | | | | | |

| Temperature: ~55–60°C | | | | | | |
|---|---|---|---|---|---|---|
| **Group 1** | temp$_{actual}$ | | | **Group 2** | temp$_{actual}$ | |
| | [borate]$_{trial 1}$ | | | | [borate]$_{trial 1}$ | |
| | [borate]$_{trial 2}$ | | | | [borate]$_{trial 2}$ | |
| | [borate]$_{trial 3}$ | | | | [borate]$_{trial 3}$ | |
| | [borate]$_{trial 4}$ | | | | [borate]$_{trial 4}$ | |
| | | | | | | |
| **Group 3** | temp$_{actual}$ | | | | | |
| | [borate]$_{trial 1}$ | | | | | |
| | [borate]$_{trial 2}$ | | | | | |
| | [borate]$_{trial 3}$ | | | | | |
| | [borate]$_{trial 4}$ | | | | | |

5. Based on your [borate] data, and that collected from the other groups, determine [borate]$_{average}$ for each temperature. Average together data from groups whose temperatures are within 0.2°C of each other (average their temperature values as well). If there are data with temperatures outside the acceptable 2°C range, consider those as separate temperature readings (i.e., if there are three groups with data collected at temperatures 44.9°C, 45.0°C, and 45.4°C, then average the data for the 44.9 and 45.0°C readings, but consider the 45.4°C data as a separate temperature reading). You should end up with a minimum of five average readings (if you can average three sets of data per temperature), or a

maximum of 15 average values (if all temperature values are far enough apart to warrant handling each group's data separately):

Temperature 1: _____ [borate]$_{average}$: _____

Temperature 2: _____ [borate]$_{average}$: _____

Temperature 3: _____ [borate]$_{average}$: _____

Temperature 4: _____ [borate]$_{average}$: _____

Temperature 5: _____ [borate]$_{average}$: _____

-------------------------------- (entries below to be used if needed) --------------------------------------

Temperature 6: _____ [borate]$_{average}$: _____

Temperature 7: _____ [borate]$_{average}$: _____

Temperature 8: _____ [borate]$_{average}$: _____

Temperature 9: _____ [borate]$_{average}$: _____

Temperature 10: _____ [borate]$_{average}$: _____

Temperature 11: _____ [borate]$_{average}$: _____

Temperature 12: _____ [borate]$_{average}$: _____

Temperature 13: _____ [borate]$_{average}$: _____

Temperature 14: _____ [borate]$_{average}$: _____

Temperature 15: _____ [borate]$_{average}$: _____

6. For each temperature, calculate $1/T$ (in units of kelvins$^{-1}$), $K_{sp}$, and ln $K_{sp}$; show examples of your work to the right and below the table.

| | Temperature (°C) | 1/T (K$^{-1}$) | $K_{sp}$ | ln $K_{sp}$ |
|---|---|---|---|---|
| 1 | | | | |
| 2 | | | | |
| 3 | | | | |
| 4 | | | | |
| 5 | | | | |
| | Use entries below if needed | | | |
| 6 | | | | |
| 7 | | | | |
| 8 | | | | |
| 9 | | | | |
| 10 | | | | |
| 11 | | | | |
| 12 | | | | |
| 13 | | | | |
| 14 | | | | |
| 15 | | | | |

**7.** Plot $\ln K_{sp}$ vs. $1/T$:

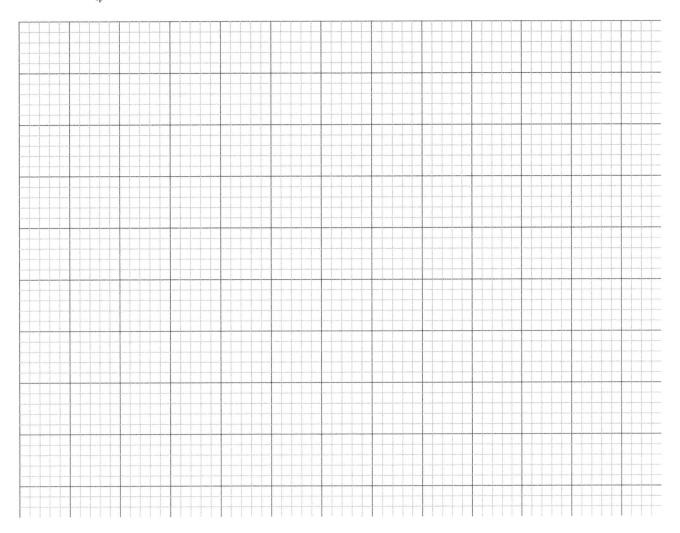

**8.** Using Excel, a graphing calculator, or the least-squares method you utilized in Laboratory 4, fit your data above to a straight line of the form:

$$\ln K_T = \left(\frac{-\Delta H_T^{\circ}}{R}\right)\left(\frac{1}{T}\right) + \left(\frac{\Delta S_T^{\circ}}{R}\right)$$

(a) Slope of your line (including units): _____

(b) $y$-intercept of your line (including units): _____

**9.** Based on #8, determine the values for $\Delta H^{\circ}$ (kJ mol$^{-1}$) and $\Delta S^{\circ}$ (J mol$^{-1}$ K$^{-1}$):

$\Delta H^{\circ}$ (kJ mol$^{-1}$): _____

$\Delta S^{\circ}$ (J mol$^{-1}$ K$^{-1}$): _____

**10.** Determine the values of $\Delta G°$ at each temperature using either Equation 3 or Equation 4. Show examples of your work below the table.

| | Temperature (°C) | $\Delta G°$ (Equation 3) | $\Delta G°$ (Equation 4) |
|---|---|---|---|
| 1 | | | |
| 2 | | | |
| 3 | | | |
| 4 | | | |
| 5 | | | |
| | Use entries below if needed | | |
| 6 | | | |
| 7 | | | |
| 8 | | | |
| 9 | | | |
| 10 | | | |
| 11 | | | |
| 12 | | | |
| 13 | | | |
| 14 | | | |
| 15 | | | |

## Reflection Questions

1. What would be the impact if you transferred a small amount of solid along with the saturated solution sample when you were generating your titration samples? Be specific about what impact (if any) this would have:

2. The literature values for enthalpy and entropy of the dissolution of borax in water are $110 \, \text{kJ mol}^{-1}$ and $380 \, \text{J mol}^{-1} \, \text{K}^{-1}$, respectively.

   (a) Determine the percent error in your experimentally determined values (#9).

   (b) What could be the possible source(s) of error leading to your results?

3. Based on your data,

   (a) Does the solubility of borax in water increase/decrease/stay the same as the temperature is increased?

   (b) Is the solvation of borax in water an exothermic or endothermic process?

   (c) Does your answer in part (a) make sense when considering your value for $\Delta H°$? Explain.

4. Clearly explain what your value of $\Delta S°$ means (i.e., is your experimentally determined value consistent with what you would predict based on Reaction 1?).

5. Was there a large degree of variability in [borate] between different groups measuring the same approximate temperature? If so, what might lead to this?

6. Suppose a solution was made with the following concentrations at 30°C: [borate] = 0.25 $M$, [Na$^+$] = 0.35 $M$.

(a) Using the results of your experiment, determine whether this system is at equilibrium or, if not, in which direction it will spontaneously proceed to reach equilibrium.

(b) Will precipitate be formed during part (a)?

## Connection

Based on your experience in this lab, draw a connection to something in your everyday life or the world around you (something not mentioned in the background section):

# 23

# Galvanic Cells and the Measurement of Cell Potential

## Objectives

- Understand the components present in a galvanic cell.
- Determine cell and half-cell potentials.
- Generate a standard reduction potential table.

## Materials Needed

- Copper, iron, zinc, tin, and silver strips, wire, or nails
- 6 to 8 test tubes (13 × 100 mm) or a 12-well plate
- 1.0 $M$ $KNO_3$
- 0.10 $M$ solutions of $CuSO_4$, $FeSO_4$, $AgNO_3$, $SnCl_2$, and $ZnSO_4$
- $CuSO_4 \cdot 5H_2O(s)$
- 10-mL graduated cylinder
- Voltmeter, multimeter, or Vernier equipped with a voltage probe
- Thick, high-quality paper towels or filter paper
- Scissors
- 3 insulated wires (separate colors) with alligator clip attachments (both ends)
- Deionized water

## Safety Precautions

- Gloves and safety goggles/glasses should be worn during this lab.
- Dispose of waste solutions in designated waste containers as indicated by your instructor.

## Background

Galvanic or voltaic cells are devices that generate electricity based on spontaneous redox reactions. They are more commonly referred to as batteries. To understand how a redox reaction is used within a galvanic cell, consider the spontaneous redox reaction involving zinc metal and an aqueous solution of copper(II) ions. As shown in the *activity series* of metals, zinc is a more active metal than copper, which means that it is more easily oxidized or loses electrons more readily. Thus, placing a piece of zinc metal into a beaker that contains an aqueous solution of copper(II) ions results in the reduction of copper(II) ions to copper metal and the oxidation of zinc metal to zinc(II) ions (Reaction 1). The transfer of electrons, the hallmark of a redox reaction, occurs as the copper(II) ions collide with the surface of the zinc metal. Because this reaction is spontaneous in the forward direction, it is necessarily not spontaneous in the reverse direction, meaning that placing copper metal into a solution of zinc(II) ions will not result in a reaction.

$$Zn(s) + Cu^{2+}(aq) \rightarrow Cu(s) + Zn^{2+}(aq) \qquad \textbf{Reaction 1}$$

In order to use this reaction (or any spontaneous redox reaction) to do electrical work, one needs access to the transferred electrons. This is not feasible if the two reactants, $Zn(s)$ and $Cu^{2+}(aq)$, are in the same container. By placing them in separate compartments and connecting the two compartments with a conducting material, typically a metal wire, the spontaneous reaction can proceed with the transferred electrons now passing through the wire as current from the zinc metal to the copper(II) ions. The compartment in which oxidation takes place, housing the zinc metal in our example, is the anode; the compartment where copper(II) ions are reduced is the cathode. Thus, electrons always travel in a galvanic cell from the anode to the cathode. This basic view of a galvanic cell is incomplete, as simply connecting two separate reactant compartments by a wire will not generate current or electron flow through the wire. If it did, the cathode compartment would become negatively charged as it receives electrons and the anode compartment would become positively charged as electrons are lost, leading to an apparatus with substantial charge separation, an energetically unfavorable scenario. (Consider the equally disfavored possibility of having separate containers of $Na^+$ and $Cl^-$ ions instead of a single container of sodium chloride—an impossibility! Nature does not support the separation of charge.) Thus, for a galvanic cell to function, an additional feature, commonly represented as a salt bridge, must be present as a connection between the two compartments. In other words, in the absence of a chemical means to balance the developing charges in both compartments as electrons are transferred, no current will flow. The salt bridge contains an ionic compound, in a mobile medium, that does not participate in the redox reaction. As electrons flow spontaneously through the wire, ions migrate through the salt bridge into the anode and cathode compartments to maintain neutrality. Cations flow from the salt bridge to the cathode compartment; anions flow from the salt bridge to the anode compartment.

A complete galvanic cell based on the spontaneous redox reaction involving $Zn(s)$ and $Cu^{2+}(aq)$ is shown in Figure 1. Zinc metal serves as the anode and is placed in a solution of $Zn^{2+}$ ions. Copper metal is the cathode and resides in a solution of $Cu^{2+}$ ions. The salt bridge allows for cation and anion migration to maintain neutrality in each compartment as electrons flow through the wire from the zinc anode to the copper cathode. A line notation is used to represent the salient features of a galvanic cell. For the cell in Figure 1, the line notation is

$$Zn(s) \mid Zn^{2+}(1.0\ M) \parallel Cu^{2+}(1.0\ M) \mid Cu(s)$$

with the contents of the anode and cathode compartments listed to the left and right, respectively, of the center lines. As current flows, the mass of the zinc electrode will decrease (zinc metal is a reactant) and $[Zn^{2+}]$ will

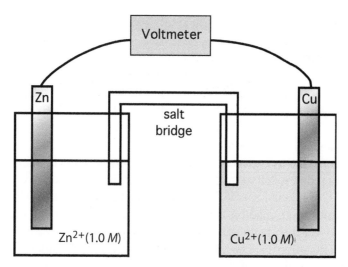

**Figure 1    A galvanic cell based on the spontaneous reaction Zn(s) + Cu²⁺(aq) → Cu(s) + Zn²⁺(aq).**

increase (a product). Likewise, the copper electrode will increase in mass (copper metal is a product) and the $[Cu^{2+}]$ will decrease. Once equilibrium is established, the reaction will be in a "finished state," and no current will flow (i.e., the galvanic cell or battery is "dead").

The voltmeter depicted in Figure 1 measures cell potential, $E_{cell}$, in volts. Cell potential is the difference in electrical potential between the anode and cathode. This potential difference reflects the relative strengths of the oxidizing agent (the species reacting in the cathode compartment) and reducing agent (the reactant in the anode compartment). For example, a stronger oxidizing agent leads to a greater "pull" of electrons through the wire, which will be reflected in a higher cell potential; a stronger reducing agent leads to a greater "push" of electrons through the wire. *Galvanic cells always have positive cell potentials* (positive cell potentials describe spontaneous redox reactions). The cell potential depends not only on the specific chemical reaction, but also on the concentrations of ions in the respective compartments and the temperature. The "half-reaction" in each compartment (anode and cathode) contributes a "half-cell" potential to the overall cell potential. With beginning solution concentrations of 1.0 $M$ in each compartment at a temperature of 25°C (typically called the standard state; while the standard state *always* refers to 1.0 $M$ concentrations for species in solution, 1 atm pressure for gaseous species, and pure solid or liquid for those physical states, a temperature of 25°C is *not* required—that is, standard state can be obtained at other temperatures), the cell potential for a galvanic cell based on the spontaneous reaction of $Zn(s)$ and $Cu^{2+}$ is 1.10 V.

|  |  | $E$ (V) |
|---|---|---|
| Oxidation half-reaction (anode): | $Zn(s) \rightarrow Zn^{2+}(aq) + 2e^-$ | 0.76 |
| Reduction half-reaction (cathode): | $Cu^{2+}(aq) + 2e^- \rightarrow Cu(s)$ | 0.34 |
| Overall redox reaction: | $Zn(s) + Cu^{2+}(aq) \rightarrow Cu(s) + Zn^{2+}(aq)$ | 1.10 |

Note that reversing any of these reactions causes a change in the sign of the cell or half-cell potential. Thus, spontaneous in one direction is not spontaneous in the reverse direction and vice versa. Further, comparing the two half-cell potentials with the half-reactions written as either reductions or oxidations verifies that $Zn(s)$ is a stronger reducing agent than $Cu(s)$ (or zinc is the more "active" metal) and that $Cu^{2+}$ is a stronger oxidizing agent than $Zn^{2+}(aq)$. Can you explain how these conclusions are reached?

## THIS LAB

The construction of various galvanic cells and the measurement of their cell potentials will be accomplished utilizing the available set of materials. The procedures are deliberately not provided for Part I of this lab. Instead, students are to work in small groups to organize their thoughts and develop appropriate experimental strategies. More direction is provided for Parts II and III. The goals for each of the three parts are as follows:

1. Construct galvanic cells, with measured cell potentials, such that a table of standard reduction potentials can be elucidated based on the following half-reactions. (*Note:* The half-cell potential for the $Zn^{2+}/Zn$ half-reaction is to be arbitrarily assigned a value of 0 V.)

| Half-reaction | $E$ (V) |
|---|---|
| $Zn^{2+}(aq) + 2e^- \rightarrow Zn(s)$ | 0* |
| $Fe^{2+}(aq) + 2e^- \rightarrow Fe(s)$ | ? |
| $Sn^{2+}(aq) + 2e^- \rightarrow Sn(s)$ | ? |
| $Ag^+(aq) + e^- \rightarrow Ag(s)$ | ? |
| $Cu^{2+}(aq) + 2e^- \rightarrow Cu(s)$ | ? |

*An arbitrary designation

2. Gain an understanding of how individual galvanic cells can be coupled *in series* to produce a working cell with higher potential. The galvanic cells to be used in this part of the lab are described by the following line notation:

$$Zn(s) \mid Zn^{2+}(0.10\ M) \parallel Cu^{2+}(0.10\ M) \mid Cu(s)$$

3. Through varying the concentration of $Cu^{2+}$ ions within cell compartments, build and understand the components in a concentration cell.

## Procedure

### I. Galvanic Cells: Development of a Table of Standard Reduction Potentials

Using the space below as well as the information given in the introduction and inspection of the available set of materials, you are to develop a logical set of procedures to accomplish the stated goal for Part I. You may wish to include a diagram(s) showing the use of the materials. To help you with organization, the procedure space is divided into two sections, *Plan* and *Modifications*. The former is your stepwise approach to addressing the goal. The latter is where you are to record any changes to your plan based on observations made during its execution. *Note:* Proper use of the voltmeter or multimeter will be demonstrated by your instructor.

**Plan**

## II.  Galvanic Cells in Series

In this part of the lab, you are to connect two galvanic cells *in series,* based on the copper and zinc half-reactions, and measure the voltage.

1. Create a galvanic cell based on the two half reactions noted above, a task that you may have already done in Part I, and record the cell potential.
2. Create a second galvanic cell, identical to that in step 1.
3. Connect the two cells *in series,* meaning that electron flow is from the anode of one cell to the cathode of the other. (Consider how batteries are commonly loaded into a flashlight, anode to cathode.) Measure the cell potential and fill in a representation of the cell in the data section below.

## III.  Development of a Concentration Cell

4. Using the Nernst equation, determine the concentration of Cu(II) needed to produce a concentration cell, in combination with the provided 0.10 $M$ Cu(II) solution, that will have a measured cell potential of 0.030 V.
5. Measure the mass of $CuSO_4 \cdot 5H_2O$ needed to prepare 10 mL of your target Cu(II) solution and prepare the solution.

6. Construct a galvanic cell using your Cu(II) solution in one compartment and the provided 0.10 *M* Cu(II) solution in the other.

7. Measure the cell potential and fill in a representation of the cell in the data section below.

## Data

### I.  Galvanic Cells: Development of a Table of Standard Reduction Potentials

1. For each cell created, provide a complete labeled representation on the accompanying diagrams (labeling of the anode and cathode compartments, composition of the anode and cathode compartments, direction of electron flow, composition and migration of the salt bridge components, and measured cell potential). Note that you may not need to create galvanic cells for all possible combinations of half-reactions to develop your table.

Cell 1:

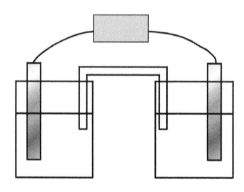

anode half-reaction (and half-cell potential):

$E(V) =$ _____

cathode half-reaction (and half-cell potential):

$E(V) =$ _____

line notation for the cell:

Cell 2:

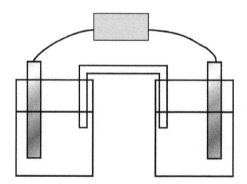

anode half-reaction (and half-cell potential):

$E(V) = $ _____

cathode half-reaction (and half-cell potential):

$E(V) = $ _____

line notation for the cell:

Cell 3:

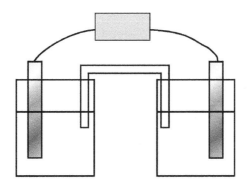

anode half-reaction (and half-cell potential):

$E(V) = $ _____

cathode half-reaction (and half-cell potential):

$E(V) = $ _____

line notation for the cell:

Cell 4:

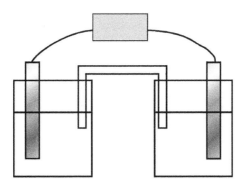

anode half-reaction (and half-cell potential):

$E(V) =$ _____

cathode half-reaction (and half-cell potential):

$E(V) =$ _____

line notation for the cell:

Cell 5:

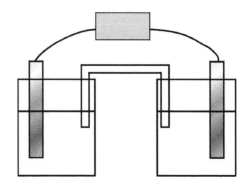

anode half-reaction (and half-cell potential):

$E(V) =$ _____

cathode half-reaction (and half-cell potential):

$E(V) =$ _____

line notation for the cell:

Cell 6:

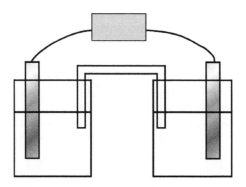

anode half-reaction (and half-cell potential):

$$E(V) = \rule{2cm}{0.4pt}$$

cathode half-reaction (and half-cell potential):

$$E(V) = \rule{2cm}{0.4pt}$$

line notation for the cell:

2. Based on your data, complete the table of standard reduction potentials below (most positive half-cell potential at the top).

| Half-reaction | E (V) |
|---|---|
| $Zn^{2+}(aq) + 2e^- \rightarrow Zn(s)$ | 0* |

*An arbitrary designation

## II. Galvanic Cells in Series

3. Measured cell potential for $Zn(s) \mid Zn^{2+}(0.10\ M) \parallel Cu^{2+}(0.10\ M) \mid Cu(s)$     $E(V) = \rule{2cm}{0.4pt}$
   (single cell, *not* connected in series)
4. On the diagram below, complete the circuit by drawing your wiring connections and provide a complete labeled representation (labeling of the anode and cathode compartments, composition of the anode and cathode compartments, direction of electron flow, composition and migration of the salt bridge components) of your final apparatus.

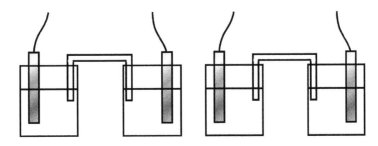

5. Measured cell potential for the cell shown in step 4.     $E(V) = \rule{2cm}{0.4pt}$

## III. Development of a Concentration Cell

**6.** Concentration ($M$) of $CuSO_4$ solution needed (show Nernst equation calculation)    _____

**7.** Mass of $CuSO_4 \cdot 5H_2O$ needed to prepare 10 mL of solution    _____

**8.** Actual mass of $CuSO_4 \cdot 5H_2O$    _____

**9.** For your concentration cell, provide a complete labeled representation on the accompanying diagram (labeling of the anode and cathode compartments, composition of the anode and cathode compartments, direction of electron flow, composition and migration of the salt bridge components and measured cell potential).

Cell 1:

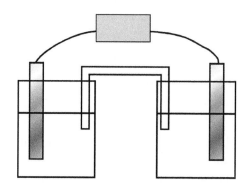

anode half-reaction:

cathode half-reaction:

line notation for the cell:

## Reflection Questions

1. What is the strongest oxidizing agent in your table of standard reduction potentials?

2. What is the strongest reducing agent in your table of standard reduction potentials?

3. Given the following relevant half-reactions present in an automobile battery

$$E° \text{ (V)}$$

$$PbSO_4(s) + 2e^- \rightarrow Pb(s) + SO_4^{2-}(aq) \qquad -0.30$$

$$PbO_2(s) + 4H^+(aq) + SO_4^{2-}(aq) + 2e^- \rightarrow PbSO_4(s) + 2H_2O(l) \qquad 1.70$$

(a) Give the overall balanced chemical equation that takes place in an automobile battery.

(b) Which half-reaction occurs at the anode?

(c) Which half-reaction occurs at the cathode?

(d) Calculate the cell potential.

(e) Based on your answer to part (d) and your experience with Part II of this experiment, explain how it is possible that an automobile battery has a cell potential of 12 V!

4. Why must one half-reaction in your table of standard reduction potentials be arbitrarily assigned a value of 0 V? In other words, why weren't all of the half-cell potentials experimentally determined?

5. The solubility of $CuSO_4 \cdot 5H_2O$ in water at 25°C is 38 g per 100 mL. Calculate the maximum cell potential that can be generated from a concentration cell based on the $Cu^{2+}/Cu$ half-reaction using a 0.10 $M$ solution of $Cu^{2+}$ in one compartment (as in this lab) and a saturated solution of $CuSO_4 \cdot 5H_2O$ in the other compartment.

6. (a) What is the cell potential for a concentration cell in which both compartments contain the same concentration of copper(II) ions?

   (b) Would the measured value for the cell potential of your concentration cell remain constant as it is used to do work? In other words, what happens to the cell potential of any galvanic cell (or battery) as it is used over time? Explain.

7. (a) Provide the line notation for *one* of the cells that you constructed in Part I of this experiment.

   (b) For the cell described in 7(a), does the mass of the anode get larger or smaller or stay the same as the cell functions?

   (c) For the cell described in 7(a), does the mass of the cathode get larger or smaller or stay the same as the cell functions?

## Connection

Based on your experience in this lab, draw a connection to something in your everyday life or the world around you (something not mentioned in the background section):

# 24

# Color Changes in Ionizing Foot Baths? (Testing Marketing Claims: A Case Study)

## Objectives

- To build and describe the components in an electrolytic cell.
- To build and describe the components in a model of a detoxifying or ionizing footbath.
- To understand the importance of scientific literacy to a society in which each individual needs to be an informed consumer.

## Materials Needed

- Deionized water
- Sodium chloride (NaCl), solid
- Steel wire
- Steel wool "Brillo" pads (*Note:* Steel contains iron alloyed or mixed with other elements.)
- Copper wire
- 6-V lantern battery
- Insulated wires with alligator clips
- pH paper
- 400-mL beaker
- Glass stirring rod
- Beral pipette

## Safety Precautions

- This experiment produces the flammable gas hydrogen. Be sure to work in the laboratory hoods.
- As a general rule, one should be careful when working with aqueous solutions and electric current. Do not put your fingers in the water while the cell is in operation.

## Background

The need for a scientifically literate society has never been more important than now, as technological advances require savvy voters and consumers willing to ask the right questions and make informed decisions on issues that span from energy to medicine to food to the environment to space exploration to even what is being taught in schools. As one example, consumer interest in fitness, nutrition, and over-the-counter health-related products and services has given marketers frequent and varied opportunities to pitch their wares via television "infomercials," the

Internet, radio talk shows, e-mail/traditional mail outlets, and so on. It is then up to the consumer to absorb and process this information using a basic understanding of scientific principles and their power to question the content and, ultimately, the sources of said information. At that point, informed decisions can be made.

One of the challenges consumers face is the use of scientific language in publicity or marketing claims that is either poorly organized, organized for confusion, or simply wrong. For example, please note the following passages taken from websites marketing detoxifying or ionizing footbaths, commercially available spa products that claim to remove toxins from your body through the pores in your feet:

"A negative ion is an atom that has lost or gained one or more electrons causing it to become negatively charged." (www.ionicoasis.com, accessed November 3, 2011)

"Our spa detoxing unit can rebalance energy meridians through the bio-charge." (http://www. lprdayspa. com/faq/index.html, accessed November 3, 2011)

"Water is an active substance, so when water comes into contact with something, the frequency of what it contacts is integrated into its own frequency structure and thus becomes its signature. When you immerse a body mass into water, the frequency is instantly added to the water as a 'memory.'" (http://andersonwellness. net/services/ion-foot-bath-detoxification, accessed November 3, 2011)

In detoxifying or ionizing footbaths, users place their feet in a basin containing a saltwater solution with subsequent connection to a power supply (an electrical outlet). The resultant electrolysis process causes changes in the composition/color of the water bath over time. Some marketers have attributed the color changes in the water to the detoxification of specific regions in the body (see Table 1).

| Table 1 Color of the footbath solution/mixture vs. body region affected. | |
|---|---|
| **Water color** | **Possible cause/condition** |
| | gall bladder, upper digestive tract |
| | kidney, bladder, urinary disorders, prostate disease |
| | arthritis, gout, joint pain |
| | blood clotting, internal bleeding |
| | smoking, fatty liver, cellular debris |
| | excessive alcohol use, liver malfunction |
| | heavy metals, diabetes |
| | insomnia, skin allergies, irregular menstruation |
| | poor digestion, headaches, constipation |

Prior to your laboratory period, you are to conduct an Internet search to gather information for a brief written summary that includes the basic design and function of a detoxifying or ionizing footbath, typical procedures for how it is used, marketing claim(s), and an example of pseudoscience babble (scientific language that is poorly organized, organized for confusion, or just plain wrong) in the form of an exact quote. Your summary is to be completed using the accompanying page. Because you will be using Internet sources, you must supply the website address and date accessed for each reference used.

# ELECTROLYSIS

Electrolysis is a process by which electrical current is used to cause chemical change. It requires a source of electrical energy that is often supplied by a battery. The same principles that were developed for the understanding of galvanic cells apply to electrolytic cells. The anode is the site of oxidation and the cathode is the site of reduction. Electrons flow from the anode to the cathode. A conducting solution is needed to allow current to flow in an electrolytic cell for reasons akin to the need for a salt bridge in a galvanic cell. The one key difference between a galvanic cell and an electrolytic cell is the requirement for a battery or source of power in the latter to drive the flow of electrons from the anode to the cathode in nonspontaneous reactions.

Electrolysis is commonly used to electroplate metals from molten or aqueous solutions of ionic compounds, allowing for layers of a metal to be deposited onto another metallic surface, or for the bulk purification of metals in their metallic form. Because reduction occurs at the cathode, electroplating of metals increases the size of the cathode over time. Electrolysis can also be used to oxidize metals, with the anode serving as the metallic source, to produce metal cations in solution. In this case, as a reactant, the anode necessarily becomes smaller as the cell runs. If the anode is a transition metal, then the solution may become colored, as many transition metal cations are colored ($Fe^{2+}$ = green; $Fe^{3+}$ = yellow to orange; $Cu^{2+}$ = blue; $Cr^{3+}$ = emerald green).

Because aqueous solutions are commonly used in electrolytic processes, including those in this laboratory, the actual oxidation and reduction half-reactions that occur at the respective electrodes will reflect the possibility that not only are solvated ions potential reactants, but water itself can participate and be either reduced to give hydrogen gas at the cathode, or oxidized to give oxygen gas at the anode. In fact, in the absence of any other reactants, electrolysis can be used to split water into the elements hydrogen and oxygen in their molecular forms [$H_2(g)$ and $O_2(g)$, respectively]. In most cases, comparing the half-reaction potentials from the table of standard reduction potentials enables us to predict the reactions that take place at each electrode, with the most easily oxidized species reacting at the anode and the most easily reduced species reacting at the cathode.

## This Lab

In this lab, your ultimate goal is to determine a method and then carry out appropriate procedures to test a marketing claim that color changes in the water of a health product, namely, a detoxifying or ionizing footbath, are due to toxins being removed through the pores of the feet. You will first be guided through the construction of an electrolytic cell to give you an understanding of the key components needed to create your own version of a footbath. In short, you will use your problem-solving skills and ample intellectual tools to "MacGyver"* a device from common, inexpensive materials. *Finally, we ask that you not share details of this lab with those outside of your lab section so that all students can have the same experience as you.*

Some key points to consider:

- Most metals are readily oxidized to form cations. If the metal is a *d*-block element or transition metal, then often the cationic form is colored (*a clue for today's experiment!*). For example, aqueous solutions of $Cu^{2+}$ ions are blue, and solutions of $Fe^{2+}$ and $Fe^{3+}$ ions are green and yellow-orange, respectively. In fact, gemstones (e.g., rubies, sapphires, emeralds, etc.) are colored because they contain transition metal cations.
- You will have a choice to use copper or steel wires in your "footbath" apparatus. Recognize the result of oxidation in either case.
- Note the chemistry that takes place at each electrode (and which battery terminal the electrode is attached to). Which involves oxidation? Which involves reduction?
- The typical detoxifying/ionizing footbath is an electrolytic cell with metal electrodes.

---

*MacGyver was a secret agent in the television series of the same name (mid-1980s) who drew on his substantial knowledge of science to solve problems using common materials around him.

# BACKGROUND: DETOXIFYING OR IONIZING FOOTBATHS

<u>Design and Function</u>

<u>Procedures for Its Use</u>

<u>Marketing Claim(s)</u>

<u>Pseudoscience Babble</u>

<u>References</u>

## Procedure, Data Collection, and Analysis

This lab will be divided into two parts. Part I will give you more detailed procedural information for constructing an electrolytic cell. You will use what you have learned in Part I to help you develop an appropriate device in Part II.

### I. Construction of an Electrolytic Cell (Figure 1)

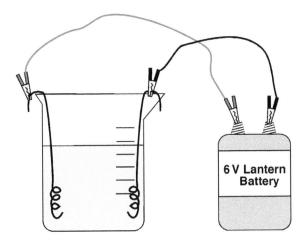

**Figure 1** **Assembled electrolytic cell.**

1. Add water to a 400-mL beaker until it is about 80% full.
2. Add a portion of sodium chloride (a scoop using a laboratory spatula should be sufficient) to the water and mix until a solution forms.
3. Measure the pH of the solution by adding a drop of the solution (using a Beral pipette) to a strip of pH paper. Do not put the pH paper directly into the solution. Record below.

   pH (acidic, neutral, or basic) _____

4. Obtain two pieces of copper wire. These will serve as your electrodes. Bend each so that the wire will hang on the side of your 400-mL beaker with the longer end extending into the aqueous solution of sodium chloride. While not a requirement, try to coil the end of the copper wire that extends into the solution.
5. Connect alligator clips from separate insulated wires to the short end of the copper wire that hangs outside of the beaker.
6. Carry your apparatus to a location within one of the laboratory hoods.
7. Connect the two insulated wires to the battery. Be sure that the two electrodes (copper wires) do not touch each other.
8. While stirring occasionally using a glass stirring rod, record observations in the following table in 3-min intervals over the course of 15 min.

| Time (min) | Observations |
|------------|--------------|
| 3 | |
| 6 | |

| Time (min) | Observations |
|---|---|
| 9 | |
| 12 | |
| 15 | |

9. Disconnect the battery and describe the appearance of the two electrodes as well as the contents of the beaker.

Anode:

Cathode:

Beaker contents:

10. Write the half-reactions that were occurring at the anode and cathode, and provide the signs on the battery ("+" or "−") that connected to the anode and cathode, respectively. (*Note:* Tables of standard reduction potentials are available in the lab.)

Anode:                                                      Sign:

Cathode:                                                    Sign:

11. Measure the pH of the solution by adding a drop of the solution (using a Beral pipette) to a strip of pH paper. Do not put the pH paper directly into the solution. Record below.

pH (acidic, neutral, or basic) _____

## II. Construction of a Detoxifying or Ionizing Footbath Model

12. Considering the remaining given materials and the components in the electrochemical cell that you created in Part I, construct an electrolytic cell that gives color changes consistent with those reported by some footbath manufacturers/distributors or health spas (see Table 1 and any references that you found). Connect the battery last and work in one of the laboratory hoods. (CAUTION: One should always be careful when working with aqueous solutions and electric current. DO NOT put your fingers in the water while the cell is in operation.) Be sure that the two electrodes do not touch each other. Be sure to consider the signs on the battery as you connect the anode and cathode. Sketch your electrolytic cell below and label *all* of the components.

**Your Electrolytic Cell (neatly sketch and label all components):**

**13.** Once assembled, record observations in the following table for the beaker contents and electrodes in 2-min intervals over the course of 12 min.

| Time (min) | Observations |
|---|---|
| 2 | |
| 4 | |
| 6 | |
| 8 | |
| 10 | |
| 12 | |

**14.** Considering the chemical species present in the cell, list all of the possible half-reactions that could have occurred at the anode and cathode. Circle the one that did occur at each electrode. You may need to consult a table of standard reduction potentials.

Anode:

Cathode:

## Reflection Questions

1. In Part I of the lab, you measured the pH of the solution before and after electrolysis. Explain any change in pH that occurred (i.e., what caused the change).

2. What was the purpose of the Brillo pad (steel wool)?

3. What was the purpose of the sodium chloride?

4. Did you use steel wire or copper wire for your apparatus in Part II? Does it make a difference? Why or why not?

   (Hint: $Cu^{2+}(aq) + 2e^- \rightarrow Cu(s)$, $E° = 0.34$ V; $Fe^{2+}(aq) + 2e^- \rightarrow Fe(s)$, $E° = -0.44$ V)

5. Were feet/body parts needed to get color changes in your footbath model?

6. Give a potential explanation, based on your data, for the observed color changes noted by some makers/distributors of detoxifying footbaths and attributed to toxins being removed from the body through the pores of the feet.

## Connection

Based on your experience in this lab, draw a connection to something in your everyday life or the world around you (something not mentioned in the background section):

# 25

# The Formation and Reactivity of Esters: A Bridge from General to Organic Chemistry

## Objectives

- Explore reactions to form and consume esters, a representative organic functional group.
- Observe the properties of esters and acids.
- Learn how to use thin layer chromatography to separate components in a mixture.

## Materials Needed

- Water
- Glacial acetic acid
- Anthranilic acid
- Salicylic acid
- Methanol
- 1-Propanol
- 1-Octanol
- Isoamyl alcohol
- Methyl salicylate
- Dichloromethane
- 5.0 $M$ Sodium hydroxide (NaOH)
- 4.0 $M$ Sulfuric acid ($H_2SO_4$)
- Concentrated sulfuric acid ($H_2SO_4$)
- Boiling chips
- 400-mL beaker
- 5 Test tubes (13- $\times$ 100-mm, 9 mL volume)
- 1 Test tube ($>$ 20 mL volume)
- 25-mL Erlenmeyer flask or beaker
- Test tube rack
- Test tube clamp
- pH paper
- Silica TLC plates with fluorescent indicator
- 3 Small vials
- TLC chamber (small jar with lid; beaker with watch glass)
- TLC spotters
- Handheld UV lamp
- Stirrer/hot plate

- Paper clip
- Ring stand and clamp
- Pipettes

## Safety Precautions

- Gloves and safety goggles/glasses should be worn during this lab.
- All reactions should be performed in a hood.
- Avoid skin contact with the organic substances in this lab (for example, methyl salicylate and salicylic acid are absorbed through the skin).
- The organic substances in this lab are flammable. Keep away from open flames.
- Avoid contact with the acids and bases in this lab. If any acid or base contacts skin, wash immediately with water and report the incident to your instructor.
- This lab involves recording the scents of volatile organic esters. The proper technique for doing so, "wafting," will be demonstrated by your instructor.
- Avoid directing the UV lamp toward anyone's eyes (including your own!).
- Dispose of waste solutions in designated waste containers as indicated by your instructor.

## Background

### CARBON

Organic chemistry is the study of carbon-based compounds. The unique chemistry of carbon allows for structures that include polymers (plastics, rubber, lubricants, adhesives, etc.), diamond, pencil "lead," medicines, dyes, textiles, paints, and the molecules of life. This remarkable structural breadth, coupled with an equally remarkable breadth of properties, would not be possible if carbon atoms did not readily form bonds with each other. The interest an element has in forming bonds with itself is called catenation and carbon is simply the catenation champion of the periodic table. One might naively surmise that catenation would be common across the elements, but in fact, no other element comes close to carbon. Catenation and an ability to form strong bonds with other atoms, primarily nitrogen and oxygen, provide the foundation for carbon atoms to serve as versatile building blocks for a vast number of molecular and polymeric structures, including those in living systems in which so many molecular parts with varied properties and function need to be present. Driven by the abundance of carbon-based sources, and an interest in both understanding living systems and the controlled manipulation of matter to produce new substances, the study of the properties of organic compounds and their chemical reactions, namely, organic chemistry, dates back to the nineteenth century and continues to be a dominant area of research today.

### FUNCTIONAL GROUPS

Because the structures of organic compounds typically contain nonpolar and relatively unreactive C—C and C—H bonds, their properties, reactivities, and identities are organized by the presence of characteristic structural subunits known as *functional groups* (Table 1). For example, replacement of one H atom in water with an alkyl or organic chain (represented by the symbol "R" in Table 1) produces a class of organic compounds called *alcohols* in which the functional group is O—H (hydroxy group). Like water, alcohols form hydrogen bonds, giving rise to properties consistent with the presence of strong intermolecular forces. Methanol and ethanol, containing one- and two-carbon R groups, respectively, are polar with relatively high boiling points for their small size because of hydrogen bonding. Following the "like dissolves like" guide that dictates solubility, both alcohols are freely miscible (soluble in all proportions) in water. However, as the carbon chain grows in length, the increase in the relative number of nonpolar C—C and C—H bonds vs. the polar O—H functionality causes the solubilities of alcohols in water to be greatly reduced. For example, just 9 mL of butanol, an alcohol containing a four-carbon R group, dissolves in 100 mL of water at 25°C, and octanol, an alcohol containing an eight-carbon R group, is immiscible with water.

In addition to sharing properties based on the O—H functional group, alcohols also undergo characteristic chemical reactions governed by the presence of the O—H bond. As part of today's lab, a variety of alcohols will be used in identical reactions with *carboxylic acids* to produce *esters* (carboxylic acids and esters are two other classes of organic compounds defined by their own distinct functional groups and properties, Table 1). The fact that different alcohols react with different carboxylic acids in a predictable manner demonstrates the link between functional group and chemical reactivity. Like alcohols, carboxylic acids maintain an O—H group and can form hydrogen bonds, while esters cannot. Esters are typically volatile substances (no hydrogen bonding) with strong fragrances, properties that will be observed in this laboratory and are most important to the food and fragrance industries. Much of synthetic organic chemistry involves the conversion of one functional group to another with subsequent introduction of new properties based on the introduced functional group. Understanding the reactions and associated mechanisms for the preparation and transformation of functional groups represents a large component of the traditional two-semester organic chemistry sequence that will be taken by nearly all of you next year. As you proceed through this lab, it is hoped that you will recognize the foundation in structure and properties that general chemistry provides to organic chemistry.

## Table 1 Functional groups.

| Class | General Formula | Structure | Functional Group |
|---|---|---|---|
| Alcohol | ROH | —Ö—H | Hydroxy group |
| Carboxylic acid | RCOOH | $\overset{\ddot{O}}{\underset{\|}{C}}$—Ö—H (C with double bond O above, bonded to O—H) | Carboxy group |
| Ester[a] | RCOOR' | —C(=O)—Ö—R' | Ester group |
| Aldehyde | RCHO | —C(=O)—H | Carbonyl group |
| Ketone | RCOR' | —C(=O)—R' | Carbonyl group |
| Amine | RNH$_2$ | —N̈(H)—H | Amino group |
| Amide | RCONH$_2$ | —C(=O)—N̈(H)—H | Amide group |

[a] R' represents a second alkyl group that may or may not be identical to the R group.

## Methyl Salicylate: A Representative Ester

From tropical rain forests to oceans to backyard gardens, nature provides an array of chemical substances of great benefit. These *natural products* are typically organic compounds, covering the full suite of functional groups, that are desired for their biological activity and structural inspirations for the synthesis of new pharmaceuticals in industrial and academic laboratories. In this experiment, you will both synthesize and react methyl salicylate, a representative ester and natural product. Methyl salicylate is found in a group of shrubs from the genus *Gaultheria* that are collectively called wintergreen. The name derives from the full-year green color of the plants and predates the more modern label "evergreen." Oil of wintergreen is 98+% methyl salicylate and is found in the leaves of wintergreen plants. Native Americans brewed tea from these leaves to provide relief from headaches, stomach aches, and fever. Methyl salicylate is also found extensively in the bark of several birch tree varieties (*Betula* genus), in particular the black or sweet birch (Figure 1).

**Figure 1** A stand of birch trees along the Elwha River, Washington. The black or sweet birch contains bark rich in methyl salicylate.

Like esters in general, methyl salicylate has a strong, pleasant aroma. Now produced synthetically, it is used for its characteristic "wintergreen" flavor and scent in candies, oral hygiene products, tobacco, soft drinks such as root beer and the aptly named birch beer, and fragrances (Figure 2). It has also found use in over-the-counter topical creams to relieve pain and stiffness. As part of this experiment, you will prepare methyl salicylate from methanol and salicylic acid to explore the properties of esters. Additionally, you will be reacting other alcohols with carboxylic acids to produce esters with different, but clearly identifiable scents. The "take home" points to this part of the lab are (1) esters are characterized by a common functional group; (2) as a group, esters are a volatile, fragrant class of organic compounds; and (3) structural changes in the ester molecule lead to different properties (i.e., different fragrances).

A curious property of wintergreen Life Savers candies called triboluminescence links to the presence of methyl salicylate. Triboluminescence is the generation of light through the mechanical breaking (crushing, rubbing, etc.) of chemical bonds. Sugar crystals are triboluminescent. Crushing a Life Saver with, for example, your teeth causes the shearing of sugar crystals with some electrons then separating from the sugar molecules. The flow of these electrons through the air as they recombine with molecules (electrical current) excites nitrogen molecules that, in turn, relax and give off ultraviolet light and faint visible light. However, in wintergreen Life Savers, the emitted light is much more intense and bluish in color because the ultraviolet light generated from the shearing of sugar crystals excites the methyl salicylate molecule, which then fluoresces in the blue region of the visible spectrum. Try it in a dark room!

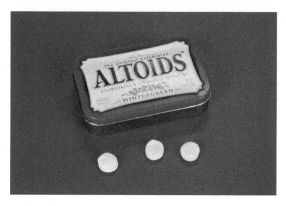

**Figure 2** Methyl salicylate is responsible for the wintergreen flavor/scent in breath mints (seen here), mouthwash, toothpaste, gum, candies, root beer, and smokeless tobacco.

Despite its ubiquitous status in the marketplace, methyl salicylate is toxic in moderate doses. Its beneficial biological activity and toxicity is tied to a structural relationship with the famed salicylic acid. Salicylic acid is one of the oldest natural remedies, serving as an anti-inflammatory drug, antipyretic (fever reducer), and general analgesic (pain reliever) for more than 6000 years until the turn of the nineteenth century. It is found naturally in fruits and vegetables and readily obtained from the bark of willow trees. Unfortunately, salicylic acid has the undesirable effect of rather pronounced mouth, throat, and stomach irritation. As such, the German dye and drug company Bayer in 1899 developed and sold a synthetic derivative of salicylic acid called acetylsalicylic acid that provided the beneficial activity of the parent natural product without the side effects. The commercial name for this invention was aspirin. You will begin today's experiment by producing salicylic acid from methyl salicylate through cleavage of the ester group with base using a reaction characteristic of esters called saponification. The toxicity of methyl salicylate is perhaps best communicated by the following: one teaspoon of methyl salicylate (5 mL) is equivalent to nearly 22 adult aspirin tablets!

## This Lab

This lab is divided into two parts based on the type of chemical reaction studied: (1) the cleavage of an ester to give a carboxylic acid and an alcohol and (2) the formation of an ester from a carboxylic acid and an alcohol. Each reaction type is the reverse of the other and involves the transformation of one or more functional groups into others.

In Part I, the ester functional group of methyl salicylate will be cleaved using a *saponification* reaction. Saponification is the cleavage or, more accurately, hydrolysis ("water-breaking") of an ester in aqueous base. It should be noted that saponification reactions have been used for several thousand years to produce soaps from triglycerides, ester-containing vegetable oils, and animal fats. In fact, the term saponification is derived from the latin "sapo," which means soap. The reaction of hydroxide with methyl salicylate will first produce salicylate (the conjugate base of salicylic acid) and methanol (Reaction 1). With hydroxide present in excess, the stoichiometry of the reaction shows that 1 mole of methyl salicylate will produce 1 mole of salicylate. After the saponification reaction is complete, the reaction mixture will be acidified to convert the salicylate into salicylic acid by the addition of acid (Reaction 2). Following filtration, the crude reaction product will be purified by recrystallization from water. In the recrystallization technique, a minimum amount of water near its boiling point is used to dissolve the product. Following cooling, salicylic acid, poorly soluble in water at room temperature, should leave the solution as a crystalline, purified substance. The observation of crystals is often a sign of a pure substance. Good technique is important here as too much water will cause the product to remain in solution upon cooling. Filtration and drying of the crystals will allow for the determination of the mass of pure salicylic acid, which, in turn, will allow for the calculation of the percent yield of the reaction. The purity of the product will then be assessed using thin layer chromatography (TLC). As described in Laboratory 15, TLC separates components in a mixture based on their relative polarities. In this lab, TLC analysis of the final product in comparison to reference samples of salicylic acid and the starting material, methyl salicylate, will be performed.

methyl salicylate + $OH^-$ $\xrightarrow{H_2O}$ salicylate + $CH_3OH$

**Reaction 1**

salicylate + $H^+$ $\xrightarrow{H_2O}$ salicylic acid

**Reaction 2**

In the second part of this lab, various esters will be prepared by combining alcohols and carboxylic acids in the presence of sulfuric acid. Sulfuric acid is a strong dehydrating agent and drives the reaction to the right. As shown specifically for the synthesis of methyl salicylate (Reaction 3), the ester-forming reactions are the reverse of the saponification reaction described in Reaction 1. In fact, a chemical equilibrium exists between an ester and its component alcohol and carboxylic acid as shown in Reaction 4. As defined in Table 1, the R and R' symbols represent alkyl chains giving rise to general formulas for carboxylic acids, alcohols, and esters. To drive the reaction toward the product side and favor formation of the desired esters, an excess of alcohol will be used for the preparation of each ester in today's lab. The produced esters have characteristic scents. Thus, you will qualitatively analyze the reaction products by recording the scent of each ester (being careful to use the demonstrated technique of "wafting" the vapor toward you).

salicylic acid + $CH_3OH$ $\xrightarrow{H^+}$ methyl salicylate + $H_2O$

**Reaction 3**

carboxylic acid + R'OH $\rightleftharpoons$ ester + $H_2O$

**Reaction 4**

## References

1. Foster S, Duke JA. *A Field Guide to Medicinal Plants and Herbs of Eastern and Central North America* (2nd edition). 1999. Houghton Mifflin Co., Wilmington, MA.
2. Leung AY, Foster S. *Encyclopedia of Common Natural Ingredients Used in Food, Drugs and Cosmetics.* 2003. Wiley-Interscience, Hoboken, NJ.

3. Chen C, Isabelle LM, Pickworth WB, Pankow JF. "Levels of Mint and Wintergreen Flavorants: Smokeless Tobacco Products vs. Confectionary Products." *Food and Chemical Toxicology* 2010; 48: 755–763.

4. Dickinson JT, Brix LB, Jensen LC. "Electron and Positive Ion Emission Accompanying Fracture of Wint-O-Green Lifesavers and Single-Crystal Sucrose." *The Journal of Physical Chemistry* 1984; 88: 1698–1701.

5. Jeffreys, D. *Aspirin. The Remarkable Story of a Wonder Drug.* 2005. Bloomsbury: New York.

6. Botma M, Colquhoun-Flannery W, Leighton S. "Laryngeal Oedema Caused by Accidental Ingestion of Oil of Wintergreen." *International Journal of Pediatric Otorhinolaryngology* 2001; 58(3): 229–232.

7. Birney DM, Starnes SD. "Parallel Combinatorial Esterification." *Journal of Chemical Education* 1999; 76: 1560–1561.

## Procedure

### I. The saponification of methyl salicylate

### A. Synthesis

1. Create a hot water bath by adding a paper clip and 300 mL of water to a 400-mL beaker. Place the beaker onto a stirrer/hot plate in a hood. Turn on the stirrer and heat the water to boiling.

2. Weigh approximately 1 g of methyl salicylate into your large test tube. You may wish to tare an assembly consisting of an empty test tube supported in a beaker. Then add the methyl salicylate directly to the test tube and record the mass to the correct number of significant figures.

3. Add a boiling chip and 6 mL of 5 $M$ NaOH to the test tube. Record any observations.

4. Using a clamp, secure the test tube in the water bath so that the entire reaction mixture is below the level of the water bath.

5. Loosely cap the test tube and heat for 25 to 30 min. To prevent the buildup of pressure inside the test tube, it is important that the cap is not fitted to the tube tightly. Record any observations.

6. Raise the test tube out of the water bath and allow the reaction mixture to cool.

7. Add 1 mL of 4 $M$ $H_2SO_4$ and stir the contents of the test tube either using a stirring rod or by gently tapping the bottom of the test tube while holding the top of the tube securely in one hand (your instructor must show you this technique). Check the pH of the reaction mixture using pH paper. Continue adding the 4 $M$ $H_2SO_4$ in 1-mL portions with subsequent complete mixing and checking of the pH until the solution is acidic. Record observations and the estimated total volume of 4 $M$ $H_2SO_4$ added.

8. Collect the solid by filtration. To collect all of the solid, wash out the residual contents in the test tube with a minimal amount of water.

9. Recrystallize the crude reaction mixture as follows. Add the solid to a clean, small Erlenmeyer flask or beaker. You may need to rinse the filter paper with a small amount (a full pipette or two) of water. Add 5 mL of water and heat the slurry to boiling on a hot plate or in your water bath. While maintaining the boiling temperature, continue to add water to the flask in small portions (1 or 2 full pipettes) until you have added just enough water to form a solution. After each addition of water, allow the mixture to return to boiling. Remove the flask from the hot plate and cool to room temperature. Place in an ice bath for 10 min of additional cooling.

10. Filter the crystalline product and oven-dry the solid for a mass determination at the end of this laboratory period. Alternatively the product may be air-dried for a mass determination at the beginning of the next laboratory period. To collect all of the solid, wash out the residual crystalline material in the flask with a minimal amount of water.

11. Describe the appearance and record the mass of the product.

12. Calculate the percent yield of the reaction.

### B. Thin Layer Chromatography

For general TLC procedures, please see Laboratory 15.

13. In three small vials, prepare three solutions by dissolving a small amount of each substance listed below (a few grains if solid, a drop if liquid) in a few drops of methanol. Label each vial.

Vial 1: Your sample of salicylic acid collected in step 10.

Vial 2: A reference sample of pure salicylic acid.

Vial 3: A reference sample of pure methyl salicylate.

14. Obtain or cut two TLC plates (plastic-backed silica plates with fluorescent indicator).
15. Place three small marks using a pencil along a line 0.5 to 1 cm from the bottom of each plate. *Note:* You do not need to draw a line, but you need to have the three pencil marks equally distant from the bottom of the plate.
16. Using a glass capillary TLC spotter, spot a sample of vial 1 on the left and center pencil marks of each TLC plate.
17. Using a new spotter, spot a sample of vial 2 on the center and right spots of one TLC plate.
18. Using a new spotter, spot a sample of vial 3 on the center and right spots of the other TLC plate.
19. Allow the spots to dry for about 1 min.
20. Add enough of a 3% methanol in dichloromethane solution to cover the bottom of the TLC chamber.
21. Place one of your TLC plates in the TLC chamber such that the solution (eluent) level remains below the marked spots on the bottom of the TLC plate. The top of the TLC plate should rest against the inside wall of the TLC chamber. Place the lid or cover on top of the TLC chamber.
22. Allow the eluent to migrate to approximately 0.5 cm from the top of the TLC plate and then remove the plate from the chamber and mark the solvent or eluent front with a pencil line.
23. Being careful not to shine the UV light in your or other's eyes, look at your TLC plate under UV light. Using a pencil, circle any observed "spots" on the TLC plate.
24. Repeat steps 20 through 22 for the other TLC plate.
25. Neatly and accurately reproduce (sketch) your developed TLCs on the provided diagrams in the Data Collection section below.

## II. The synthesis of esters

The following procedure will be used for the preparation of each ester from the combination of the following alcohols and carboxylic acids. Use a laboratory fume hood and the hot water bath prepared in step 1.

- methanol and salicylic acid
- methanol and anthranilic acid

26. Place 100 to 200 mg of carboxylic acid in a 13- × 100-cm test tube.
27. Add 10 drops of the alcohol and gently mix the contents of the tube.
28. Add one drop of concentrated sulfuric acid. Place the tube in a water bath, heated to boiling, for 5 min.
29. Remove the test tube from the water bath and place in a test tube rack. Add 3 to 4 mL of water.
30. Agitate the contents of the test tube by gently tapping the bottom of the tube while holding the top of the tube securely in one hand. Your instructor will demonstrate this technique.
31. Wave your hand across the top of the test tube to direct the vapor toward your nose. This technique is known as "wafting" and should be done with caution. Do not place the test tube directly under your nose. Observe the scent and record its description in the data table provided.

The following procedure will be used for the preparation of each ester from the combination of the following alcohols and acetic acid. Use a laboratory fume hood and the hot water bath prepared in step 1.

- octanol and acetic acid
- isoamyl alcohol and acetic acid
- *n*-propanol and acetic acid

32. Place 2 drops of glacial acetic acid into a 13- × 100-cm test tube.
33. Add 10 drops of the alcohol and gently mix the contents of the tube.
34. Add one drop of concentrated sulfuric acid. Place the tube in a water bath, heated to boiling, for 5 min.

35. Remove the test tube from the water bath and place in a test tube rack. Add 3 to 4 mL of water.
36. Agitate the contents of the test tube by gently tapping the bottom of the tube while holding the top of the tube securely in one hand. Your instructor will demonstrate this technique.
37. Wave your hand across the top of the test tube to direct the vapor toward your nose. This technique is known as "wafting" and should be done with caution. Do not place the test tube directly under your nose. Observe the scent and record its description in the data table provided.

## Data Collection

### I. The saponification of methyl salicylate

## SYNTHESIS

1. Mass of methyl salicylate (step 2):

    _____

2. Observations (step 3):

3. Observations (step 5):

4. Observations (step 7):        Estimated total volume of added 4 $M$ $H_2SO_4$ _____

5. Appearance of recrystallized final product (step 11):

6. Mass of final product, salicylic acid (step 11):

    _____

7. TLC sketches (label each lane in each chromatogram with the substances spotted). Be sure to include a line for the solvent/eluent front and the spots observed under UV light.

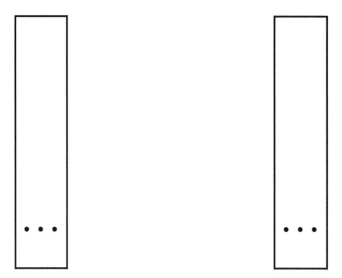

final product vs. *methyl salicylate* reference          final product vs. *salicylic acid* reference

## II.  The synthesis of esters

8. Record the scents of the esters produced from the following combinations of alcohols and carboxylic acids.

| | Acetic Acid | Anthranilic Acid | Salicylic Acid |
|---|---|---|---|
| Methanol | | | |
| Isoamyl alcohol | | | |
| Octanol | | | |
| *n*-Propanol | | | |

# *Analysis*

## I.  The saponification of methyl salicylate

9. Determine the limiting reactant in your preparation of salicylic acid (show calculation).

**10.** Percent yield for the preparation of salicylic acid, step 12 (show calculation):

_____

**11.** Explain how the TLC data supports the identity and purity of your salicylic acid sample.

**12.** Considering your TLC data, which is more polar, methyl salicylate or salicylic acid? Explain.

**13.** Draw the structures of the following esters that were prepared in Part II of this experiment.
   (a) Octyl acetate (octyl refers to an eight-carbon alkyl group or chain of carbon atoms)

   (b) *n*-Propyl acetate (propyl refers to a three carbon alkyl group or chain of carbon atoms)

## Reflection Questions

1. Provide the Lewis structure for salicylic acid, including all bonds and lone pairs of electrons. Circle the most acidic proton.

2. As you have learned in this experiment, esters are typically volatile and fragrant. Would you expect esters to be more or less volatile than carboxylic acids? Why?

3. Provide an explanation as to why methyl salicylate is a liquid, but salicylic acid is a solid.

4. TLC was used to help identify and assess the purity of your final product in Part I of this experiment. Suggest another method that might allow you to identify your product as pure salicylic acid.

5. In the synthesis of the five esters (Part II of this experiment), concentrated sulfuric acid was added and described as a dehydrating agent. Considering Reaction 4 in the introduction, what would be the purpose of adding a dehydrating agent in the preparation of esters?

6. Methyl salicylate and salicylic acid are natural products. Using any resources necessary (books, Internet, etc.), find another natural product (your choice) and report the following information about it.

    (a) Name:

    (b) Structure:

    (c) Natural source (plant/animal species, location, etc.):

    (d) Biological activity and/or use (anticancer drug, analgesic, etc.):

    (e) Source of information:

Based on your experience in this lab, draw a connection to something in your everyday life or the world around you (something not mentioned in the background section):

# Photo Credits

## Lab 1
**Lab 1.1:** © Pixtalage/Fotostock RF.

## Lab 2
**Lab 2.1:** © Pixtalage/Fotostock RF; **Lab 2.3(left):** © The McGraw-Hill Companies, Inc. Jill Braaten, photographer; **Lab 2.3(right):** © The McGraw-Hill Companies, Inc. Terry Wild Studio.

## Lab 3
**Lab 3.6:** Gregg R. Dieckmann and John W. Sibert.

## Lab 5
**Lab 5.4:** Gregg R. Dieckmann and John W. Sibert.

## Lab 7
**Lab 7.1:** © Stockbyte/Getty RF; **Lab 7.2(both):** © The McGraw-Hill Companies, Inc. Jacques Cornell, photographer; **Lab 7.3:** © Corbis RF; **p. 79:** © The McGraw-Hill Companies, Inc. Louis Rosenstock, photographer.

## Lab 9
**Lab 9.1:** © The McGraw-Hill Companies, Inc. Stephen Frisch, photographer.

## Lab 10
**Lab 10.1:** © The McGraw-Hill Companies, Inc. Charles D. Winters, photographer; **Lab 10.2(both):** © The McGraw-Hill Companies, Inc. Charles D. Winters, photographer; **Lab 10.4:** © The McGraw-Hill Companies, Inc. Eric Misko, Elite Images Photography.

## Lab 11
**Lab 11.1:** © Getty RF.

## Lab 13
**Lab 13.5(both):** Gregg R. Dieckmann and John W. Sibert.

## Lab 14
**Page 160:** © Getty RF.

## Lab 15
**Lab 15.1:** © NHPA/Photoshot RF; **Lab 15.2:** © Keith Eng.

## Lab 25
**Lab 25.1:** © Getty RF; **Lab 25.2:** © The McGraw-Hill Companies, Inc. Matt Meadows, photographer.